I0035103

FACTORY AND SOURCING CHECKLISTS

CERM Academy Series On Enterprise Risk Management

Greg Hutchins PE CERM
503.233.1012 or 800.COMPETE
GregH@QualityPlusEngineering
CERMAcademy.com

HOW TO ORDER:

Cost is $69.00 per copy. Quantity discounts are available from the publisher
Quality Plus Engineering, LLC
4052 NE Couch St.
Portland, OR 97232
USA
503.233.1012 (USA) or 800.COMPETE or 800.266.7383
GregH@europa.com

On your letterhead, please include information concerning the intended use of the books and the number of books you wish to purchase.

©2018 CERM ACADEMY/QUALITY PLUS ENGINEERING

This publication is designed to provide accurate and authoritative information in regard to the subject matter covered. It is sold with the understanding that the publisher is not engaged in rendering legal, accounting, or other professional services. If legal advice or other expert assistance is required, the service of a competent professional person should be sought.

TABLE OF CONTENTS

SECTION	PAGE
Continuous Improvement	7
Corrective Action	9
Corrective Action Procedures	10
Corrective Action Requests	11
Customer Sales	12
Customer Satisfaction	13
Customer Selling/Marketing	17
Customer Service Personnel	18
Customer-Stakeholder Relationships	20
Customer-Stakeholder Requirements	21
Customer-Supplier Communications	22
Design and Development	25
Design Change Control	26
Design Configuration Controls	27
Design Drawings	28
Design New Products	31
Design Product Development	33
Design Quality	35
Design Quality Inputs	36
Design Quality Outputs	37
Design Quality Review	38
Design Review	40
Design Specifications	42
Document Controls	43
Energy Management	45
Environmental Emissions	47
Environmental Hazardous Waste	48
Environmental Policy	49
Environmental Product Testing	50
Environmental Recovery	51
Environmental Recycling	52

Environmental Site Selection	53
Environmental Waste	54
Financial Investment	55
Financial Management	56
Financial/Physical Resources	57
Human Resource Development	58
Human Resource Management	61
Human Resource Organization	64
Human Rights	66
Information Technology	67
Information Technology Control	69
Information Technology Development	70
Inspection and Testing of Supplied Products	72
Inspection Procedures	74
Legal and Regulatory Compliance	78
Lessons Learned Policy	79
Material Disposition Procedures	81
Material Distribution	83
Material Labeling	84
Material Movement and Handling System	86
Material Packaging	88
Material Quality	90
Material Shipping Procedures	91
Material Storage Procedures	93
Material Transportation	95
Measurement and Test Equipment	96
Measurement and Test Procedures	98
Measurement Equipment Control	100
Nonconformance Control	101
Preventive Action Requests	102
Project Closeout	103
Production and Manufacturing	104
Production Documentation	106
Process Monitoring, Control, and Improvement	108
Product Reliability Testing	112
Product Testing	114

Production/Delivery of Services	115
Production Environment	116
Product and Workplace Safety	117
Production/Manufacturing	120
Production Materials	122
Production Planning	123
Production Preventive Maintenance	125
Production Quality	126
Production Quality Review	127
Production Review and Control Plans	128
Project Management Resources	130
Purchasing Quality	131
Quality Auditing	132
Quality Control	134
Quality Costs	136
Quality Documentation	138
Quality Goals and Objectives	141
Quality Information and Analysis	144
Quality Management Principles	148
Quality Management System	149
Quality Management System Goals	150
Quality Management System Reviews	151
Quality Management/Assurance Reviews	153
Quality Control Reviews	154
Quality Management Team	155
Quality Manager	156
Quality Manual	157
Quality Measurement	159
Quality Organization	161
Quality Planning (Assurance)	162
Quality Planning Control	163
Quality Plans (General)	164
Quality Results	166
Quality Workplace	167
Risk Management – Emergency Planning	169
Sales/Marketing Strategy	170

6   Factory and Sourcing Checklists

Strategic Management	171
Suppliers/Contractor Quality	172
Supplier Partnering	173
Supplier Training	175

CONTINUOUS IMPROVEMENT

1) Is a continuous improvement ethic adopted companywide?
2) Is a continuous improvement ethic adopted supplier wide?
3) Is prevention pursued companywide?
4) Are problems prioritized in terms of importance or value?
5) Is companywide quality improvement a company goal?
6) Is there a quality improvement plan at:
   a) Corporate level?
   b) Business unit level?
   c) Plant level?
   d) Process level?
7) Is supplier wide quality improvement a company goal?
8) Are companywide benchmarks established?
9) Are performance targets identified in company operations?
10) Is quality improvement actively led and promoted by top management?
11) Is first-level supervision supportive of the improvement effort?
12) Do line personnel pursue quality improvement daily?
13) Do management, supervision, and line personnel understand area processes?
14) Is lowering total costs a major element of the improvement effort?
15) Is process variation monitored in:
   a) Manufacturing?
   b) Administrative operations?
   c) Service operations?
   d) Professional areas?
   e) Supplier operations?
16) Are internal process controls established companywide?
17) Are these process concepts universally understood?
   a) Variation
   b) Process control
   c) Process capability
   d) Process improvement
18) Is there a structured effort to identify and eliminate assignable causes of variation?

19) Is there a structured effort to identify and eliminate common causes of variation? Is variation around determined targets systematically minimized?
20) Are there improvement efforts with hourly and salary employees?
21) Do quality improvement teams meet weekly or monthly?
22) Are quality improvement efforts communicated to the auditee's organization through:
    a) Charts?
    b) Progress reports?
    c) Meeting minutes?
    d) Cost/benefit analyses?
23) Are the goals, objectives, benefits, responsibilities, costs, and timelines of continuous improvement efforts reviewed?
24) Are improvement records maintained and monitored?

CORRECTIVE ACTION

1) Is there a formal corrective action program?
2) Is it used to resolve the following:
   a) Process control and capability checks?
   b) Customer complaints?
   c) Nonconforming supplied material?
   d) Field failures?
   e) Machine capability?
3) Are corrective action procedures developed for:
   a) Regulatory and environmental complaints?
   b) Raw materials?
   c) Process systems?
   d) Products?
   e) Packaging and labeling?
   f) Special projects?
   g) Facilities?
4) Does the program provide for prompt detection of inferior quality?
5) Is the information used to improve systems, processes, and products?
6) Are analyses made to identify deficiency trends and patterns?
7) Is corrective action taken to arrest unfavorable trends before deficiencies occur? Are symptoms and root causes both eliminated?
8) Is there a post audit to validate or verify elimination?
9) Does corrective action extend to supplier's products?
10) Is corrective action taken in response to customer data?
11) Are data analysis and product examination conducted on scrap or rework to determine extent and causes of nonconformances?
12) Is the effectiveness, efficiency, and economy of corrective action monitored?
13) Does corrective action also focus on continuous improvement?
14) Does the quality group have the authority to require process improvement?
15) Are improvements measured?
16) Are users consulted about the necessity of improvement?

CORRECTIVE ACTION PROCEDURES

1) Does manufacturing provide the following information to the customer:
   a) Quality cost reports?
   b) Problem solving results?
   c) Corrective action results?
   d) Improvement results?
   e) Reaction to concerns?
2) Analysis of root causes?
3) Is there an effective corrective action system?
4) Is corrective action audited after implementation?
5) If quality problems occur, are they solved quickly and efficiently?
6) Is corrective action used to select and improve manufacturing performance?
7) Is manufacturing-wide continuous improvement pursued?
8) When deficiencies occur, does manufacturing identify the symptom, identify the root cause, and eliminate both the symptom and root cause?
9) Does manufacturing provide internal customers with a written correction action plan with implementation timing?
10) Is the resolution of a problem audited?
11) Do parts and processes require regulatory approval?
12) If yes, are site auditors stationed to observe testing?
13) Are remote manufacturing facilities audited?
14) Are remote manufacturing facilities audited by third-party auditors?
15) Does quality management lead and co-ordinate audits?

CORRECTIVE ACTION REQUESTS

1) Has the Quality Manager established a process for reducing or eliminating design, production, operations, and supplier nonconformances?
2) Is corrective action required of nonconformances resulting from?
   a) Stakeholder complaints
   b) Nonconformance reports
   c) Management review results
   d) Quality assurance review reports
   e) Quality control review reports
   f) Data analysis results
3) Do corrective action requests:
   a) Identify the nature of the nonconformance?
   b) Determine causes of nonconformances?
   c) Evaluate actions to ensure that nonconformances do not recur?
   d) Recommend actions to ensure that nonconformances do not recur?
   e) Record the results of actions taken?
   f) Review corrective actions taken are effective and recorded?
4) Are corrective action requests (CAR) and preventive action requests (PAR) prioritized using the following criteria:
   a) Level 3: Critical? This nonconformance level results in an environmental, safety, or operational health, ethical, and/or security hazards?
   b) Level 2: Major? This nonconformance level negatively affects design, production, or operational integrity or performance?
   c) Level 1: Minor? This nonconformance level is a minor noncompliance with a specification or standard?
5) Does the Quality Management Team:
   a) Review the nature and cause of the nonconformance and assign the priority level?
   b) Determine the time required for CAR closure?
   c) Review recommended actions?
   d) Review the effectiveness of the CARs and/or PARs upon completion?
   e) Maintain a current log of outstanding CARs and PARs?

CUSTOMER SALES

1) Is information provided to customers for proper use, storage, and disposal of the product?
2) Are distributors and retailers trained in how to promote and use the product?
3) Do customer service or help desk personnel know products sufficiently well to provide customer service?
4) Does advertising accurately reflect product characteristics?
5) Is advertising effective?
6) Is there a process to evaluate and reconsider current advertising effectiveness?
7) Are target customer segments identified?
8) Are the customer requirements of each segment identified?
9) Are plans developed to satisfy these market segments with products that satisfy these segments?
10) Are customer product attributes prioritized?
11) Is there a formal procedure to deploy these within the organization and into the customer base?
12) Are there customer metrics?
13) Are customer complaints or suggestions tracked to identify patterns?
14) Is the success of the above strategies monitored and revised as necessary?
15) Are employees incented for achieving or surpassing customer satisfaction targets?
16) Are there policies and procedures for evaluating current designs?

CUSTOMER SATISFACTION

1) Auditing External Customer Requirements
2) Are external customers profiled in terms of:
   a) Niches?
   b) Characteristics?
   c) Demographics?
   d) Psychographics?
   e) Requirements?
3) Are trends and levels of customer satisfaction monitored?
4) Are current and future customer requirements monitored?
5) Are historical customer concerns identified?
6) Are product characteristics that affect form, fit, function, durability, governmental regulations, and safety identified?
7) Are plans established to meet customer concerns and requirements?
8) Is customer satisfaction compared against:
   a) Principal competitors?
   b) Industry averages?
   c) Industry leaders?
   d) World leaders?
9) To identify customer requirements, does the auditee use:
   a) Customer questionnaires?
   b) Customer interviews?
   c) Focus groups?
   d) Service level analysis?
   e) Quality studies?
   f) Field failure reports?
   g) Customer complaint analysis?
   h) Corrective action analysis?
   i) Warranty records?
   j) Quality costs?
   k) Customer returns?
   l) Quality function deployment?
   m) Benchmarking?
   n) Dealer visit reports?
   o) Market test reports?
   p) Supplier quality reports?

q) Preproduction builds?

r) Product development/validation reports?

10) Are customer requirements communicated:

a) Through product development?

b) Through the product lifecycle?

11) Is there a formal process for communicating customer needs, wants, expectations to the auditee?

12) Are the following data and information collected:

a) Performance information?

b) Complaints?

c) Cost of quality?

d) Customer satisfaction?

e) Competitor's performance?

13) Are customer requirements and expectations deployed throughout the auditee's organization?

14) Is there a formal process for communicating customer requirements to suppliers?

15) Are the following indicators monitored:

a) Complaints?

b) Claims?

c) Refunds?

d) Recalls?

e) Returns?

f) Repeat services?

g) Litigation?

h) Replacements?

i) Downgrades?

j) Repairs?

k) Warranty costs?

l) Warranty work?

m) Field failures?

n) Correction costs?

o) Auditing External Customer Complaints?

16) Is there an active effort to resolve customer complaints for quality improvement?

17) Are complaints resolved promptly?

18) Is there a system for classifying customer complaints?

19) Are the most severe or critical customer complaints handled first?

20) Is a complaint report generated for:
    a) Performance?
    b) Packaging?
    c) Labeling?
    d) Delivery?
    e) Ordering?
    f) Service?
    g) Pricing?
    h) Technical specifications?
21) If yes, does the report communicate:
    a) Customer name?
    b) Customer location?
    c) Date of complaint?
    d) Customer PO number?
    e) Product name?
    f) Manufacturing location?
    g) Manufacturing date?
    h) Shipping location?
    i) Shipping date?
    j) Lot number?
    k) Date of receipt?
    l) Date of resolution?
    m) Failed sample?
    n) Description of failure?
    o) Test results?
    p) Conclusions?
    q) Recommendations?
    r) Corrective action?
22) Are complaints studied to determine patterns or trends?
23) Is the complaint report sent to the department or individuals who can diagnose and correct the problem?
24) Is the customer sent a resolution report?
25) Does the resolution report:
    a) Assign responsibility for report?
    b) Diagnose problem?
    c) Identify symptoms?
    d) Identify root cause?
    e) Identify risks?

f) Recommend corrective action?

g) Identify resources for corrective action?

h) Date corrective action?

i) Identify person responsible for corrective action?

26) Is there customer follow-up after a purchase to evaluate customer satisfaction?

27) Is the customer relationship periodically reviewed for improvement?

CUSTOMER SELLING/MARKETING

1) Is product marketing and selling planned and managed throughout the product life cycle?

2) For each market segment and product, are the following addressed, monitored, and managed?
   a) Pricing
   b) Placement
   c) Promotion
   d) Branding
   e) Merchandising
   f) Advertising
   g) Communication

3) Are specified target customers and their requirements specially identified?

4) Is there a long term as well as short-term sales forecasts?

5) Is there a plan and metrics to ensure sales are met?

6) Have sale plans been develop as well as terms for sales?

7) Is there a formal and computerized process to communicate orders into the production and delivery process?

8) Is there a JIT or formal forecast system?

9) Is there an ERP system?

CUSTOMER SERVICE PERSONNEL

1) Are the following addressed for customer service personnel:
   a) Selection factors?
   b) Career path?
   c) Special training?
   d) Empowerment?
   e) Problem solving?
   f) Attitude assessment?
   g) Recognition and reward?
   h) Attrition?
2) Are customer service personnel trained in:
   a) Product knowledge?
   b) Listening to customers?
   c) Soliciting customer feedback?
   d) Handling special problems?
   e) Skills in customer retention?
3) Are there employee participation programs, such as:
   a) Quality circles?
   b) Improvement teams?
   c) Corrective action teams?
4) Are employee recognition programs in place?
5) Are internal customers identified for most administrative, service, and manufacturing operations?
6) Do suppliers satisfy internal customers?
7) Is internal customer satisfaction measured?
8) Is there a feedback mechanism to the internal customers and suppliers?
9) Are internal performance and quality targets defined and understood?
10) Is the health of customers considered during product design or development?
11) Are there clear instructions on how to use, dispose of products?
12) Are product risk or injuries considered in product development?
13) Are products simple to use?
14) Are product controls and marking easy to understand?
15) Are the appropriate regulatory health warnings on the product, machine, tags, and containers?

16) Are the following considered in product packaging?
   a) Language
   b) Instructions
   c) Weight of the product
   d) Mode of transportation
   e) Accessibility
   f) Product integrity
17) Are racial, cultural, ethnic, and sexual considerations addressed during product development and marketing?

CUSTOMER-STAKEHOLDER RELATIONSHIPS

1) Is there a formal process for identifying and managing stakeholder requirements?
2) Is there a formal process for communicating with organizational stakeholders?
3) Is there a formal public relations imitative?
4) Is the board of Directors satisfied with the level and quality of information received from management?
5) Have there been any legal or ethical lapses?
6) Are these issued corrected and prevented from occurring?
7) Are the following corporate quality issues managed?
   a) Process improvement
   b) Risk management
   c) Service and product quality
   d) Cost of quality
   e) Cycle time reduction
   f) Performance and productivity
   g) Benchmark improvement
8) Are operational baselines and benchmarks established?
9) Are these benchmarks used for improvement?
10) Have 'best in class' benchmarks been established?
11) Is there a system for process improvement, such as Six Sigma?
12) Are business processes reengineered or reinvented periodically?
13) I there a commit to total quality management?
14) Is TQM managed over the product lifecycle?

CUSTOMER-STAKEHOLDER REQUIREMENTS

1) Does the Quality Manager or a management representative have the authority and responsibility to ensure critical stakeholders are identified?
2) Are stakeholder requirements are clearly understood and written down?
3) Are stakeholder requirements are converted into achievable quality objectives and metrics?
4) Is stakeholder satisfaction is monitored and reported to senior corporate management with the aim of increasing stakeholder confidence in quality?
5) Are stakeholder requirements addressed and interfaces managed production?

CUSTOMER-SUPPLIER COMMUNICATIONS

1) Are supplier-partners consulted early in the product development cycle regarding:
   a) Specifications?
   b) Quantities?
   c) Capabilities?
   d) Delivery dates?
   e) Costs?
   f) Technology?
   g) Contracts?
2) Do products and services satisfy customer requirements?
3) Are customer's quality requirements spelled out to suppliers?
4) Are supplier's quality requirements spelled out to subcontractors?
5) Are quality requirements tied to a purchase order?
6) Do the customer and key supplier partners jointly develop product specifications?
7) Are proprietary products purchased?
8) Are there alternate suppliers of proprietary products?
9) Is there a procedure for securing proprietary or sensitive data?
10) Are proprietary products thoroughly tested?
11) Is all critical information included in the purchase order sent to suppliers?
12) Are suppliers responsible for understanding customer requirements and expectations?
13) Are suppliers responsible for questioning inconsistencies or omissions?
14) Do suppliers submit a formal feasibility statement to the customer upon initial quotation?
15) Are control plans for critical and major quality characteristics approved by the customer's engineering department?
16) Are the latest engineering drawings, engineering specifications, and material specifications sent to suppliers?
17) Are dimensions on engineering drawings identified for quick reference to initial sample documents?
18) Are written policies/procedures/ instructions available and sent to suppliers for:
   a) Handling material?

b) Shipping material?

c) Storing material?

d) Conducting inspection?

e) Conducting testing?

f) Securing test equipment?

g) Assessing Communication of Material?

h) Specifications?

19) Have customer or suppliers identified the appropriate material specifications, whether:

a) International Organization for standardization (ISO)?

b) National standards (ANSI, DIN, BSI)?

c) Industry standards (SAE, ASTM, AGA, API, etc.)?

d) Company standards?

20) Are specifications available for all incoming raw material?

21) Does quality management coordinate the preparation of specifications?

22) Does the customer share design, performance, delivery and service data with supplier partners?

23) Are suppliers required to maintain a current specification file with readable drawings and specifications?

24) Do specifications indicate quality levels?

25) Are current, accurate and complete quality documentation sent to suppliers?

26) Do suppliers inform the customer of problems in quality, delivery, and specifications?

27) Do suppliers check only critical and major specifications?

28) Do suppliers have facilities to perform measurements specified on specifications?

29) Are specifications realistic?

30) Assessing Sharing of Engineering Drawings and Communication via Contract?

31) Are all dimensions and quality characteristics specified on the engineering drawing?

32) Is there any special notation or symbols that have to be explained to the supplier?

33) Do engineering prints use geometric dimensional tolerancing (GDT)?

34) Do suppliers provide SPC charts on designated quality characteristics with each shipment of the material?

35) Are customer business and legal requirements spelled out in a purchase order and/or contract?
36) Does the supplier conduct contract reviews?
37) Does the customer sit down and ex plain contract requirements to suppliers?
38) If a disagreement or a difference occurs between customer and supplier, is there a means to arbitrate or mediate the dispute without resorting to litigation?

DESIGN AND DEVELOPMENT

1) Are environmental conditions factors considered in product design and development?
2) Are products redesigned or improved to be environmentally friendly?
3) Is ecological thinking used in the product life cycle from creation, specification, production, use, reuse and ultimate disposition?
4) Is there research to determine disposal or reuse of obsolescent products?
5) Is new production machinery adaptable to improvement in technology?
6) Are parts designed for multiple uses and for reusable?
7) Do parts and products conform to ASTM, ANSI, ISO, or applicable standards?
8) Are parts suppliers ISO 9000 compliant?
9) Are parts designed to be customizable?
10) Is a common platform used in the design of new products?
11) Is growth in production capacity planed?
12) Are multiple suppliers chosen in the event of an unforeseen event?

DESIGN CHANGE CONTROL

1) Does the Quality Manager review, approve, and record authorized design changes?
2) Does the result of the design changes result in subsequent follow up actions, which can be recorded?
3) Does the Quality Manager or quality representative determine the effect of design changes on:
   a) Interaction between design and production?
   b) Interaction between designers and sub designers?
   c) Interaction and coordination among stakeholders?

DESIGN CONFIGURATION CONTROLS

1) Is a person or team designated for maintaining and updating the parts lists and bills of material?
2) Are drawings and bills of material maintained on a computer?
3) Is there a formal process of controlling drawing changes?
4) Are engineering drawing changes priori tiled in terms of importance?
5) Can anyone initiate an engineering drawing change?
6) Are engineering changes reviewed and approved by the following:
   a) Supervising engineer?
   b) Quality assurance?
   c) Multidisciplinary team?
   d) Key suppliers?
   e) Engineering control board (ECB)?
7) Is there a procedure for withdrawing superseded drawings?

DESIGN DRAWINGS

1) Are the following drawings used:
   a) Layout?
   b) Assembly?
   c) Detail?
   d) Exploded/isometrics?
2) Are the following considered in the bill of materials?
   a) Parts identified by part number and name?
   b) Parts numbered to correspond to drawing location?
   c) Proper quantity required?
   d) Specification information?
3) Are drawings and specification revisions controlled?
4) Are all critical or control quality attributes identified on engineering drawings?
5) Are quality attributes prioritized and identified on drawings?
6) Are drawings properly detailed?
7) Are dimensions that affect fit, form, function, durability, governmental regulations and safety identified?
8) Is geometric dimensional tolerancing (GDT) used?
9) Are reference dimensions identified to minimize inspection layout time?
10) Are sufficient control points and datum surfaces identified to design functional gages?
11) Are process dimensions compatible with accepted manufacturing standards?
12) Are any requirements specified on engineering drawings that cannot be evaluated using known inspection measurement techniques?
13) Are bills of material complete and accurate?
14) Are special production processes identified on drawings?
15) Are unusual conditions or instructions specified in notes?
16) Is distribution of drawings controlled?
17) Are supplied materials made from proprietary designs?
18) Are on-site reviews conducted?
19) Do engineering documentation or specifications prescribe specific tests?
20) Are the following detailed in drawings:
   a) Dimensions?

   b) Tolerances?

   c) Notes?

   d) Special instructions?

   e) Bills of material?

   f) Special processes?

   g) Materials of construction?

   h) Interfaces?

   i) Assemblies/subassemblies/components?

   j) Arrangement?

   k) Instrumentation?

   l) Electrical?

   m) Software/hardware/firmware?

21) Are drawings reviewed through drawing cycle?

22) Are the following systems of drawing control used:

   a) Numbering systems?

   b) Distribution?

   c) Use authorization?

23) Do the following review and approve drawings:

   a) Draftsperson?

   b) Engineers?

   c) Supervising engineer?

   d) Multidisciplinary team?

   e) Suppliers?

24) Does the control plan specify the Sample size?

25) Are measured values recorded for all drawing specifications?

26) Are all dimensions, quality characteristics specified on engineering drawings?

27) Is there special notation or symbols that have to be explained?

28) Does the auditee check all drawing specifications?

29) Does the auditee have facilities to perform all the measurements?

30) If not, does the contractor performing the measurements need to be certified?

31) Is this material identified on the shipping papers and/or container?

32) Following a deviation request, is there a procedure requiring corrective action to eliminate need for deviation requests?

33) Is appearance specified on engineering drawings?

34) Does appearance consider features such as:

   a) Color?

    b) Grain?

    c) Finish?

    d) Lighting?

35) Do appearance features require customer engineering's approval?

36) Does the auditee retain a master layout sample?

37) If yes, does the sample have an approval date?

38) Are blanket statements of conformance to quality specifications, standards, and engineering drawings acceptable?

39) Is there a formal procedure for control ling nonconforming or dated engineering records, drawings, and specifications?

40) Does the procedure address:

    a) Criticality of the nonconformance?

    b) Segregation of nonconformance?

    c) Reporting of nonconformance?

    d) Documenting nonconformance?

    e) Disposition of nonconformance?

    f)   Corrective action?

DESIGN NEW PRODUCTS

1) Are new products and services developed according to an overall plan?
2) Are new products and services developed based on a project model?
3) Are new product and services developed on time and on budget?
4) Are customer requirements incorporated into new product/services plans and concepts?
5) Is there an identifiable product life cycle?
6) Is there a plan to manage the product life cycle?
7) Are product/service development target established, including:
   a) Quality?
   b) Cost?
   c) Delivery?
   d) Service?
   e) Technology?
8) Is leading technology, including Web enabling incorporated into the new product/service?
9) Is a prototype built and tested prior to deployment?
10) Is the following addressed in the prototype stage?
    a) Value engineer
    b) Constructability
    c) Service specifications
    d) Concurrent engineering
    e) Manufacturability
11) Are patents and other intellectual properties developed and protected?
12) Is there a strategy or plan to refine existing products or services?
13) Is this being considered at specified periods in the Product Life Cycle?
14) Is there a plan to improve quality, reliability, maintainability, or service throughout the product life cycle?
15) Is there a plan to test the efficiency and effectiveness of new or revised products or services?
16) Is there a plan for preparing the product or service for production?
17) Is there a prototype production process?
18) Have suppliers been selected?
19) Has all production equipment been designed and tested?
20) Is the product/service development process monitored, controlled, and measured?

DESIGN PRODUCT DEVELOPMENT

1) Is there a formal process or procedure detailing product design and development?
2) Is there an engineering manual and does it address:
   a) Customer requirements?
   b) Product development management?
   c) Engineering procedure?
   d) Design requirements?
   e) Drawing control?
   f) Specification control?
   g) Engineering drawing change?
   h) Calculation review?
   i) Software/hardware/firmware review?
   j) Testing?
   k) Records retention?
3) Are these controls used in design calculations:
   a) Design review and approval?
   b) Engineering calculations page numbering and dating?
   c) Assumptions review?
   d) Units of measure checking?
   e) Formula review?
   f) Calculation crosschecking?
4) Are technical abilities sufficient to develop products?
5) Are cost/benefit studies conducted to design and develop products?
6) Are the following considered in developing a product:
   a) Product simplification?
   b) Product optimization?
   c) Failure mode analysis?
   d) Tolerance studies?
   e) Design for assembly?
   f) Finite analysis?
   g) Solid modeling?
   h) Simulation?
   i) Reliability analysis?
   j) Maintainability analysis?
   k) Cost reduction?

l) Capability?

m) Safety?

n) Manufacturability?

o) Computer simulation and other

p) Testing?

q) Selection and developing of suppliers?

r) Hiring and training personnel?

s) Software and hardware?

t) Instructions, manuals, and documentation?

u) Packing design?

v) Shipping, handling, and storage?

w) Test and measurement equipment?

x) Field support?

7) Is an interdisciplinary approach used in developing products?

8) If yes, are the following groups consulted and used:

a) Product engineering?

b) Manufacturing engineering?

c) Design engineering?

d) Manufacturing/production personnel?

e) Industrial engineering?

f) Quality?

g) Purchasing?

h) Costing?

i) Production control?

j) Legal?

k) Suppliers?

l) Field service?

m) External customers?

n) Cost accounting?

o) Field support?

9) Are design calculations reviewed and approved?

10) Is new product development planned?

11) Are key suppliers incorporated into product development?

12) Is product development formally managed?

13) Are project management methods used, such as:

a) Critical path method?

b) PERT?

14) Are the following addressed in the product development project:

a) Project objectives and outcome?
b) Project milestones?
c) Assignments and accountabilities?
d) Interrelationships and relationships?
e) Organization?
f) Budgets and resource availability?
g) Schedules?
h) Meeting and communication loops?
i) Progress review?
j) Cost estimates?
k) Planning?

15) Is there a formal design review process?

DESIGN QUALITY

1) Has the Quality Manager planned the assurance and control of design quality and the control of subsequent production quality throughout the lifecycle of a product or project?

2) Do designs proceed beyond go/no-go gates (30, 60, 90, 100% submittals) without the input of all specified parties and the written approval of the designated project or the Quality Manager?

3) Does the organization prepare design quality plans, which at a minimum include:

   a) Identify go/no-go review gates (30, 60, 90, 100% submittals) of the design review process?

   b) Identify required review, verification, and validation activities at each critical design review gate (30, 60, 90, 100% submittals)

   c) Conduct risk assessment at the critical design gates?

   d) Identify responsibilities and authorities for design review, verification, and validation?

   e) Interface and coordinate with designers, sub designers to resolve critical design issues?

   f)   Interface with production managers on constructability reviews?

   g) Work with critical stakeholders to root-cause resolve quality deficiencies and variances?

4) Do critical suppliers prepare design plans, which at a minimum include:

   a) Identify go/no-go review gates (30, 60, 90, 100% submittals) of the design review process?

   b) Identify required review, verification, and validation activities at each critical design review gate (30, 60, 90, 100% submittals)?

   c) Conduct risk assessment at the critical design gates?

   d) Identify responsibilities and authorities for design review, verification, and validation?

   e) Interface and coordinate with designers, sub designers to resolve critical design issues?

   f)   Interface with production managers on constructability reviews?

   g) Work with critical stakeholders to root-cause resolve quality deficiencies and variances?

DESIGN QUALITY INPUTS

1) Are critical stakeholder requirements identified and incorporated in a 'statement of design requirements'?
2) Are conflicting or ambiguous requirements surfaced, clarified, and resolved at the earliest review gate (submittal) of the design process?
3) Does the Quality Manager:
   a) Review design criteria and assumptions?
   b) Obtain design requirements from stakeholders?
   c) Review applicable regulatory and legal requirements?
   d) Review environmental requirements?
   e) Review lessons learned from previous designs?
   f) Collect and gather any other quality requirements essential for design consistency and production conformance?
4) If designs are contracted to architect/engineering firms, are the above activities completed with external designers and sub designers?

DESIGN QUALITY OUTPUTS

1) Are the outputs of the design process recorded in a format that enables verification at defined review and acceptance gates (30, 60, 90, 100% submittals)?

2) Does design output:

   a) Satisfy requirements?

   b) Contain or make reference to design and production quality acceptance criteria?

   c) Define design quality characteristics that are essential to safety, constructability, operability, maintainability, and risk?

   d) Secure all approvals before progressing to the next design gate?

DESIGN QUALITY REVIEW

1) Does the Quality Manager conduct design reviews at defined intervals and gates (30, 60, 90, 100% submittals) during the design process to verify conformance to requirements and consistent quality?
2) Can the Quality Manager request that contractor designers and sub designers submit quality plans detailing how they are complying with customer quality requirements?
3) At the defined gates (30, 60, 90, 100% submittals) are systematic design reviews conducted to:
   a) Evaluate capability to fulfill stakeholder quality requirements?
   b) Identify adequacy of design?
   c) Identify problems areas, potential hazards, and potential design risks?
   d) Identify interface and coordination issues with production management?
   e) Identify interface areas, which may facilitate manufacturing or production?
   f) Produce design and production solutions to identified quality problems?
   g) Ensure quality criteria, specifications have been addressed in the contract documents?
   h) Ensure design process is consistent and thorough?
4) Do critical reviewers/stakeholders participate at design gate reviews (30, 60, 90, 100% submittals)?
5) Do design reviews identify and document deficiencies in order to initiate corrective or preventive actions?
6) Design Quality Verification and Validation?
7) Are designs verified to ensure that quality requirements are satisfied?
8) Are contractor designs verified and detailed in their submitted quality plans?
9) Does design verification depend on project criticality including:
   a) Peer design review?
   b) Calculation review?
   c) Drawing control review?
   d) Design assumption review?
   e) Sub designer control?

f) Self and independent checking?

g) Alternative design calculations, computer simulation if required?

h) Design assurance audits and corrective actions if required?

10) Are designs validated to confirm that designs can be constructed with a minimum of change orders?

11) Does design validation depend on project criticality including:

a) Design for constructability?

b) Reviews by stakeholders?

c) Value analysis?

12) Is validation defined, planned, and completed prior to forwarding designs to production?

13) Are validation results and subsequent follow up action recorded?

DESIGN REVIEW

1) Are modified product designs formally reviewed?
2) Is design review a team effort?
3) If applicable, does the design team encompass all disciplines?
4) Does a team follow design through product development and product life cycles?
5) If yes, does it incorporate feedback from different disciplines?
6) Is there a standing committee responsible for design reviews?
7) Does the design review committee include representatives from?
   a) Design engineering?
   b) Product engineering?
   c) Quality?
   d) Manufacturing?
   e) Accounting?
   f) Purchasing?
   g) Suppliers?
   h) Marketing?
   i) Customers?
8) Does design review address:
   a) Specification reviews?
   b) Preliminary designs?
   c) Drafting checks?
   d) Calculation reviews?
   e) Interdisciplinary reviews?
   f) Selection of standard parts and materials?
   g) Use of commonly accepted standards?
   h) Personnel training?
   i) Manufacturability?
   j) Supplier input?
   k) Interface and input with all parties?
   l) Periodic reports to top management?
   m) New market/customer requirements?
   n) Costs/benefits?
   o) Constraints?
   p) New features?
   q) Key design assumptions?

    r) Alternative design analysis?

    s) Selected approach and supporting rationale?

    t) Anticipated risks and contingencies?

9) Are key design calculations checked?

10) Do suppliers have an opportunity for design input?

DESIGN SPECIFICATIONS

1) Are external customer requirements identified?
2) Are internal customer requirements identified?
3) Are external customer requirements communicated to engineering through a formal mechanism, such as quality function deployment?
4) Are engineering prints developed from customer requirements?
5) Do specifications identify a product's quality characteristic?
6) Do specifications state the desired level of quality for a product quality characteristic?
7) Is there a formal documented specification system?
8) Is responsibility for maintaining the system clearly defined?
9) Are specifications periodically reviewed so they reflect process capabilities?
10) Is there a formal mechanism to approve specification changes?
11) Do users understand specifications?
12) Are specifications developed for:
    a) Raw material?
    b) Processed material?
    c) Supplied material?
    d) Fabricated material?
    e) Processes?
    f) Products?
13) Is there a procedure for preparing, approving, and controlling specifications?

DOCUMENT CONTROL

1) Has senior management established programmatic policies required to maintain and secure institutional memory?
2) Does the Quality Manager ensure that these policies result in project-level procedures for controlling design, production, quality, and other critical documents?
3) Do these policies and procedures conform to document control system and policies?
4) Has senior management authorized adequate resources to maintain control and if required digitize Quality Management System documentation?
5) Does quality documentation consist of three levels of documentation?
   a) Quality policies
   b) Quality procedures
   c) Quality control instructions
6) Do document control processes ensure that:
   a) Design, production, quality, and other critical stakeholder documents are approved for adequacy prior to release?
   b) Critical documents are reviewed and updated as necessary and reapproved?
   c) Controlled versions of relevant quality documents are available at appropriate locations where quality activities are conducted?
   d) Quality documents are controlled to the prevent unintended use?
   e) Obsolete quality documents that are retained for legal or knowledge
7) Does a master list of controlled quality documents?
   a) Identify the current revision status of documents
   b) Preclude the use of invalid and obsolete documents
   c) Reside in the quality organization
8) Has the Quality Manager identified critical design prints, production information, and quality information required to determine quality conformance?
9) Are measurement and verification uniquely identified and traceable to the source?
10) Is traceability required of critical quality documents?
11) Are critical quality documents legible, readily identifiable, and retrievable?

12) Are quality documents uniquely identified and recorded?

ENERGY MANAGEMENT

Background
1) What percentage of overall costs is the cost of energy?
2) At what rate is energy costs increasing?
3) Is there a formal program to track and to manage energy costs?
4) Is there an energy conservation team?
5) Doe the energy team consist of stakeholders from production, planning, purchasing, engineering and other functions?
6) Do some employee's job descriptions include participation in this group?
7) Are the energy project teams' responsibilities scalable to the energy intensiveness of the organization?
8) Are employees designated to review energy use in specific areas of the organization?
9) Is there a formal ecological team within the organization?
10) Are social and ecological goals identified and pursued?
11) Do the organization and its suppliers foster ecological innovation and long-term sustainability?
12) Are human rights issues addressed in sourcing decisions?

Research
1) Is research and development conducted on relevant energy conserving technologies or products?
2) Are there employee education workshops on energy conservation?
3) Are employee suggestions solicited to manage energy costs?

Energy audits
1) Are energy audits conducts?
2) Are the organizational areas, which consume the most electrical or heat energy identified?
3) Are energy costs identified?
4) Is the total cost of energy purchased calculated, monitored, and tracked?
5) Are final savings from conservation monitored?
6) Are financial savings and energy conservation a priority with senior management?

7) Is energy audit data collected and reviewed on a regular basis?
8) Are energy conservation ideas routinely collected from employees?
9) Are new technologies identified for reducing energy consumption?

Conservation Measures
1) Are plans developed for managing energy consumption, such as installing temperature regulators?
2) Is energy monitoring equipment installed?
3) Is cogeneration used to recover heat?
4) Is water temperature monitored and regulated on a continuous basis?
5) Are areas of high-energy consumption identified?
6) Is energy usage per product or per area identified?
7) Does high-energy consumption result in air, water, or soil pollution?
8) Is energy use managed?
9) Is energy a criterion considered in product design?
10) Is energy optimized?
11) Is energy conservation routinely communicated to employees?
12) Are energy saving practices collected and communicated throughout the organization?

ENVIRONMENTAL EMISSIONS

1) Is the organization compliant with EPA and other environmental regulations?
2) If there is an infraction, are effective corrective and preventive actions taken?
3) Do the same environmental noncompliances recur?
4) Is waste production and emission regularly or continuously monitored?
5) Is measuring equipment controlled?
6) Is there a specific team to control and mitigate hazardous conditions?
7) Do employees have sufficient technical knowledge of process production waste and emissions?
8) Are individual by products of manufacturing specifically identified?
9) Are the following identified and prioritized?
    a) Toxicity
    b) Lifetime
    c) Medical hazards
    d) Safety hazards
10) Are plans developed for disposing of waste?
11) Are disposal plans evaluated for effectiveness?
12) Are procedures written for disposing of hazardous materials and controlling emissions?
13) Are the following controlled?
    a) Emissions
    b) Noise
    c) Toxic waste
    d) Water pollution
14) Are all emissions identified from buildings and factories?
15) Are the origins of the emissions identified?
16) Are the consequences and risks of each emission identified and prioritized in terms of risk?
17) Are measures implemented to eliminate high-risk emissions?

ENVIRONMENTAL HAZARDOUS WASTE

1) Are hazardous materials identified and identified according to regulations?
2) Are potential carcinogens identified?
3) Are contingency plans developed in the event of an environmental accident?
4) Are automatic monitors and other warning systems installed in the event of an emergency?
5) Is there a postproduction or 'end of pipe' environmental monitoring and protection equipment?
6) Are scrubbers and other equipotent installed to reduce hazardous emissions from air stacks?
7) Is there a policy to adopt best practices to reduce environmental emissions?
8) Are inefficient and ineffective pollution equipment replaced with efficient ones?
9) Are alternative means of generating energy used?
10) Are hermetic processes adopted for dangerous dust producing substances such as asbestos?
11) Are harmful substances responsibly recycled, reprocessed, recovered, or disposed according to regulations?
12) Are manufacturing processes redesigned to eliminate harmful substances?
13) Are manufacturing or distribution processes redesigned to be more efficient, effective, or safe?

ENVIRONMENTAL POLICY

1) Are environmental issues addressed in contracts with suppliers and sub suppliers?
2) Are political and human rights issues addressed in supply contracts?
3) Do policies and procedures detail acceptable human rights sourcing?
4) Are some national or regional manufacturers deemed unacceptable for sourcing?
5) Are suppliers and sub suppliers induced to meet new human rights and child employment standards?
6) Are timetables and project plans developed to improve offshore working conditions?
7) Are sustainability and ecological advancement part of the organizational strategy and practices?
8) Are these policies deployed throughout the organization?
9) Are less damaging production processes identified?
10) Are suppliers selected and induced to developed integrated process of wastewater reclamation, erosion control, and habitat restoration?
11) Are suppliers selected who practice biocontrol, use no artificial fertilizers, or pesticides?
12) Are suppliers selected who use alternative fuels, energy, and transportation?
13) Is there an independent system for independently assuring ecological innovation and sustainability?
14) Is there an independent system for independently assuring that suppliers meet working conditions and ecological contractual requirements?
15) Are independent labs used to evaluate potentially dangerous materials?

ENVIRONMENTAL PRODUCT TESTING

1) Are computer modeling or simulation used to design and test products?
2) Are toxic materials identified during product design and development?
3) Is there a plan to identify hazardous materials?
4) Are contingency plans developed if there is a mishap?
5) Are harmful preservatives or trace contaminants identified?
6) Are endangered or threatened species addressed at any point?
7) Are endangered/threatened plant species, deforestation/defoliation, considered in product development, design, sourcing, distribution, and marketing?

ENVIRONMENTAL RECOVERY

1) Is there a formal recover and recycle program?
2) Are converters installed to improve raw material yields or to reduce emissions?
3) Are procedures and practices used to minimize cooling water use?
4) Is automatic measuring and monitoring equipment installed in the facilities?
5) Are metals, acids and solvents recovered?
6) Is there a process for recycling, rinsing, and cleaning water?
7) Are formal processes used to recycle, recalculate or reuse products or convert them into other products?
8) Is industrial swage water separated and reprocessed?
9) Are hazardous materials identified?
10) Are hazardous materials replaced with less hazardous ones?
11) Is low sulfur coal used rather than high surfer coal?
12) Are cadmium coatings replaced with zinc coatings?
13) Are PCB transformers replaced?

ENVIRONMENTAL RECYCLING

1) Are there formal, written recycling policies and procedures?
2) Are costs of switching to recycled products been considered?
3) Is recycling considered in product design and development?
4) Is there a procedure to standardize components?
5) Are non-biodegradable materials avoided?
6) Are raw materials recovered and waste reduced during manufacturing processes?
7) Are recycling bins conveniently placed and labeled?
8) Are customers offered rebates for recycling or returning products?
9) Are employees educated on the importance of recycling?
10) Are suppliers educated and induced on recycling products?
11) Are there procedures for handling, transporting, and storing high-risk materials?
12) Is there a policy or procedures to use biodegradable substances in manufacturing processes?

ENVIRONMENTAL SITE SELECTION

1) Are there formal policies and procedures for sting plants and facilities?
2) Are the following considered in siting facilities?
   a) Geological faults
   b) Floods
   c) Radon gas
   d) Security
   e) Cost
   f) Permitting
   g) Suppliers
3) Are building designed for the well-being of users?
4) Is there an asbestos removal program?
5) Are their formal policies and procedures for reducing?
   a) Noise
   b) Dust
   c) Air quality
   d) Fumes
   e) Chemicals
   f) Toxic materials
   g) Electromagnetic radiation
6) Are employee complaints and illnesses tracked to look for patterns?
7) Are external building materials and internal materials tracked for health risks?
8) Is asbestos removed from all facilities?
9) Is landscape maintained according to regulations and internal procedures?
10) Are neighboring residents surveyed to determine feelings?
11) Are open spaces and habitat restoration pursued?

ENVIRONMENTAL WASTE

1) Are waste management plans implemented?
2) Are there procedures for reducing and eliminating waste?
3) Are procedures developed to reduce waste?
4) Are alternative methods used for reducing production waste?
5) Are waste products used as material input for another production process?
6) Are there disposal methods for waste?
7) Do disposal methods comply with all applicable regulations?
8) Are deposals methods ranked in terms of their effectiveness, ease of implementation, and financial costs?
9) Are the most effectiveness means pursued to reduce and eliminate waste?
10) Are task forces or teams designated to reduce and eliminate waste?
11) Are centrally located receptacles used for storing waste and by products?
12) Are collection points installed for used oil, solvents acids, dyes, etc.?
13) Is toxic waste stored in well-ventilated areas away from employees and heat?
14) Are facility containers used for waste paper, plastics, glass, aluminum, and metals?

FINANCIAL INVESTMENT

1) Do investment policies and procedures prescribe property, plant, and equipment purchase?
2) Is there a procedure for cost-benefit decision making for evaluating production investment?
3) Does the organization comply with Federal, state, and local purchasing regulations?
4) Doe the organization invest in employee training?
5) Does the organization incent employees through stock options plans and compensation benefits?
6) Are tax credits and benefits used properly by the organization?

FINANCIAL MANAGEMENT

1) Are customers invoiced on time?
2) Are accounts payable and accounts receivable managed?
3) Is there a formal process for monitoring and correcting billing problems?
4) Is there a process for resolving billing inquires and problems?
5) Is there a formal after sales service to resolve complaints and other issues?
6) Is there a formal process of handling?
   a) Customer inquiries
   b) Warranties
   c) Claims
7) Is there a formal process for handling warranties and claims?
8) Is the Web used to enhance or enable the customer-manufacturer and customer-supplier process?

FINANCIAL/PHYSICAL RESOURCES

1) Are financial resources audited and managed?
   a) Budgets
   b) Resource allocations
   c) Capital outflows
   d) Cash flows
   e) Financial risks
   f) Accounting
   g) Accounts payable
   h) Accounts receivable
   i) Collections
   j) Credits
   k) Closing the books
   l) Retiree information
   m) Travel and entertainment expenses
2) Is financial information reported according to FASB and GAAP?
3) Are there sufficient internal financial and administrative controls to minimize risks?
4) Are internal audits conducted?
5) Is the tax function managed for tax compliance?
6) Is there an overall tax strategy?
7) Is IT effective employed in this area?
8) Are tax controversies resolved quickly?
9) Are tax records current and accurate?
10) Have the external auditors reported any deficiencies or findings?
11) Is there an effective internal auditing program?
12) Is internal auditing respected and value process?
13) Is there a risk management and mitigation program?
14) Are the following risks mitigated?
   a) Physical resources
   b) Capital planning
   c) Fixed assets
   d) Facilities

HUMAN RESOURCE DEVELOPMENT

1) Does the auditee identify external and internal customer needs and develop training to satisfy these needs?
2) Is there a continuous training and development effort?
3) Is everyone in the auditee's organization trained in quality principles and methodology?
4) Are job classifications, descriptions, and accountabilities documented, maintained, and updated for all personnel?
5) Is training measured against results in the workplace?
6) Are new employees introduced to quality culture, organization expectations, and needs?
7) Are employees updated periodically in new developments?
8) Does the auditee assess internal training needs by asking internal customers?
9) Is training required or offered to:
   a) Management?
   b) First line supervision?
   c) Technical staff?
   d) Line workers?
   e) Administrative workers?
   f) Field personnel?
   g) Suppliers?
   h) Service personnel?
10) Is continuous training offered to all personnel?
11) Are the following quality topics covered in courses?
   a) Customer service
   b) Continuous improvement
   c) Simultaneous engineering
   d) Quality statistics
   e) Statistical Process Control (SPC)
   f) Flowcharting
   g) Check sheet
   h) Pareto analysis
   i) Fishbone diagrams
   j) Poka yoke
   k) Histogram

l) Scatter diagrams

m) Run diagrams

n) Benchmarking

o) Cause and effect diagram

p) Quality costing

q) Design of experiments

r) Quality function deployment

s) Team building

t) Time management

u) Workflow analysis

v) Problem solving

w) Project management

x) Blueprint reading

y) Geometric dimensional tolerancing

z) Brainstorming

aa) Value engineering

bb) Workflow analysis

cc) Quality auditing

dd) Quality function deployment

ee) Designing for manufacturability

ff) Testing and measurement

12) Are records maintained for:

a) Certification?

b) Training?

c) Retraining?

13) Is training specific to the company, process, and product?

14) Are the following considered when training personnel?

a) Product knowledge

b) Safety

c) Attitude

d) Skills

e) Knowledge

f) Operational procedures

g) Equipment knowledge

h) Customer relations

15) Is training offered through:

a) On the job training (OJT)?

b) Seminars?

   c) Customer plant visits?

   d) Supplier plant visits?

   e) Schools?

   f) Role playing?

   g) Simulation?

   h) Videotapes?

   i) Consultants?

   j) Professional societies?

16) Are individual training records maintained and updated?

17) Are internal areas or people targeted for training due to:

   a) Knowledge obsolescence?

   b) New technologies?

   c) Expansion of duties?

18) Does everyone attend training sessions?

19) Is each job documented and are instructions available?

20) Are new training techniques implemented?

21) Are personnel instructed in all equipment and machines?

22) Is a train-the-trainer approach used?

23) Does the auditee have a tuition reimbursement program?

24) Is training conducted with the unions?

25) Are personnel cross-trained in upstream and downstream operations?

26) Are personnel trained similarly in:

   a) Use of equipment?

   b) Responding to customer requests?

HUMAN RESOURCE MANAGEMENT

1) Is senior management actively involved in:
   a) Setting quality goals and objectives?
   b) Planning quality?
   c) Reviewing quality performance?
   d) Communicating with employees?
   e) Recognizing employee contributions?
   f) Participating in quality management teams?
   g) Identifying competitor's quality advantages?
   h) Meeting with customers?
   i) Meeting with suppliers?
   j) Establishing quality benchmarks?

2) Is senior management actively involved in building quality values with:
   a) International groups?
   b) National groups?
   c) Community agencies?
   d) Trade groups?
   e) Business organizations?
   f) Educational groups?
   g) Health care organizations?
   h) Standards making groups?
   i) Government organizations?

3) Is there a process for communicating quality values?

4) Does the company try to determine how well quality values are adopted throughout the auditee's organization by using any of the following methods?
   a) Surveys
   b) Interviews
   c) Focus groups

5) Is cooperation promoted among employees across organizational levels and functions?

6) Are business units encouraged and guided to develop quality plans and goals?

7) Does the auditee share its quality leadership and knowledge on:
   a) Health and safety?
   b) Environmental protection?

c) Business ethics?

d) Waste management?

8) Is top management committed and involved in the quality program?

9) Has the auditee applied for the Malcolm Baldrige Award? If yes, what was the composite and functional area scores?

10) Does the company have a customer who has or is applying for the Malcolm Baldrige? If yes, does the customer re quire application?

11) Is a minimum score required for continued business?

12) Does general management actively direct and participate in the quality program?

a) Auditing Human Resource Management's

b) Commitment to Quality of Hiring and Training

c) Employee Advancement

13) Are human resources important to the auditee?

14) Are human resource plans developed?

15) Do these plans address:

a) Training and development?

b) Hiring?

c) Employee involvement?

d) Suggestions systems?

e) Empowerment?

f) Recognition and rewards?

16) Do these plans incorporate:

a) Internal partnering?

b) Supplier partnering?

c) Labor management cooperation?

d) Recognition systems?

e) Increasing employee accountabilities?

f) Increasing employee authorities?

g) Education and training initiatives?

17) Does the auditee have a succession plan for key executives?

18) Is there an expressed hiring and advancement policy?

19) Do interviewers outline clear and realistic quality expectations to prospective employees?

20) Do advertisements and interviewers follow equal opportunity regulations?

21) Are supervisors and team members brought into the selection and hiring process?

22) Are reference checks conducted?
23) Is there a formal employee recognition and performance measurement system?
24) Is quality improvement a part of every one's appraisal system?
25) Is quality performance routinely evaluated?
26) Is quality performance measured against written and known criteria instead of subjective criteria?
27) Is team performance recognized?
28) Is individual performance rewarded and recognized?
29) Are teams or individuals informed of performance through appraisals and counseled on how to improve?
30) Are performance reviews conducted periodically and objectively?
31) Is first-level supervision trained to appraise, correct, and reward performance?
32) Does first-level supervision consistently direct daily quality efforts?
33) Does auditee's organizational chart list all key positions and accountabilities?
34) Are all positions staffed?
35) Are administrative and clerical personnel overworked?
36) Is the average tenure for each position short?
37) Is there high turnover or absenteeism?
38) Are employees consulted prior to being transferred or before changes are made in their work environment?
39) Are suggestions and information re viewed and pursued?
40) Are suggestions acknowledged or rewarded?
41) Are these used to communicate or so licit ideas or information:
    a) General business forums?
    b) Suggestion boxes?
    c) Open door policy?
    d) Problem-solving meetings?
    e) Quality circles or other participatory group meetings?
    f) Newsletters?
42) Does management and supervision listen to and implement suggestions?

HUMAN RESOURCE ORGANIZATION

1) Is there a plan to develop, manage, educate, and retain human resources?
2) Are human resources strategies aligned with corporate strategies?
3) Are present and future human resource requirements understood?
4) Is there a plan to bring online the appropriate human resources as demand requires?
5) Is the cost of human resource training and hiring kept?
6) Does the Human Resources department have a strategic role?
7) Are human resource requirements understood for each new product development and deployment process?
8) Is there a formal process of cascading human resource tactics throughout the organization?
9) Are human resources requirements defined and aligned with processes?
10) Are work requirements and people competencies defined?
11) Is the deployment of people management?
12) Is the person's lifecycle managed?
13) Is work redesigned along process lines?
14) Is the following employee lifecycle managed?
    a) Recruiting
    b) Selection
    c) Correcting
15) Is training and development a standardized process?
16) Are employee direction and goals aligned with the corporate vision and mission?
17) Is there an orientation program for new employees?
18) Is employee performance measured and appropriate feedback delivered?
19) Is there a 360-degree evaluation program?
20) Are self-managing team used throughout the organization?
21) Is team performance measured?
22) Are employee performance, reward and recognition measured?
23) Is work measured for market value and internal equity?
24) Is there a process for base and variable compensation?
25) Is worker retention measured?

26) Is employee morale measured and improved?
   a) Labor management relations
   b) Employee involvement
   c) Health and safety
   d) Internal communications
   e) Employee benefits
   f) EEO and diversity
   g) Collective bargaining
   h) Human Resources Information Systems

HUMAN RIGHTS

1) Does the organization have policies and procedures addressing:
   a) Workplace violence?
   b) Cultural diversity?
   c) Disability?
   d) Social responsibility?
   e) Human rights?
2) Are there formal policies and procedures for addressing how these are correcting and preventing recurring noncompliance?
3) Is the organization sensitive to local architecture, food, language, culture in advertising, and selling strategies?
4) Is the organization sensitive to racial, class, ethnic, religious, or sensual stereotyping?
5) Does the organization promote ecological and sustainable initiatives?
6) Is the organization sensitive to cross cultural differences?
7) Does the organization adopt and comply with global labor, safety, and product standards?
8) Does the organization promote ISO 9000 or ISO 14000 registration internal as well as with suppliers?
9) Are internal and external suppliers audited for compliance with labor and safety standards?
10) Are new materials and labor standards addressed in contracts with suppliers?
11) Does the organization have plans in the eventuality of economic boycotts of business operations in offshore countries?
12) Does the organization have policies regarding discrimination in employment, promotion, race, etc.?
13) Are environmentally degrading products and processes discontinued for the protection of the environment?
14) Are offshore operations or suppliers audited for?
   a) Compliance with contracts
   b) Compliance to local laws or regulations
   c) Sustainability
   d) Cultural fit

INFORMATION TECHNOLOGY

1) Is there a formal system for managing information technology?
2) Are IT process and project defined and management corporate wide?
3) Are IT initiatives aligned with:
   a) Business plan?
   b) Business strategies?
   c) Business tactics?
   d) Business processes?
4) Is there a plan for enterprise system architecture?
5) Are enterprise data standards developed?
6) Are there IT quality and process standard and controls?
7) Are IT user needs and requirements defined thorough a formal capturing method?
8) Is there an IT enterprise support structure?
9) Is there a formal structure for selection IT technologies?
10) Are data a system life cycle defined and developed?
11) Is there a formal process for testing, evaluating, and deploying enterprise IT support systems?
12) Are systems security and control deployed?
13) Have systems security strategies and level been developed and deployed?
14) Is the information storage and retrieval process defined and managed?
15) Is there a formal database system to?
   a) Acquire and collect information
   b) Store information
   c) Modify update information
   d) Enable retried information
   e) Delete information
   f) Secure information
16) Are IT facilities and networks managed and secured, including:
   a) Centralized facilities?
   b) Distributed facilities?
   c) Network facilities?
17) Are the following managed, controlled, and measured:
   a) Libraries?
   b) Business records?

c) External communication systems?

d) Internal communications systems?

e) Publications?

18) Are IT processes periodically audited?

19) Are corrective and preventive actions implemented as a result of the audits?

INFORMATION TECHNOLOGY CONTROL

1) Does the Quality Manager identify and capture critical quality information necessary to manage programmatic, project, and activity consistency?

2) Is critical quality information identified, collected, and secured to develop organizational institutional knowledge and memory?

3) Does the Quality Management System ensure the availability, adequacy, fitness for use, storage, protection, confidentiality, and retrieval of critical quality information and documentation?

INFORMATION TECHNOLOGY DEVELOPMENT

1) Is there a formal and consistent procedure for software development and testing?
2) Does software comply with any of the following standards?
   a) Department of Defense (DOD)
   b) Institute of Electrical and Electronic Engineers (IEEE)
   c) Military standards (MIL-STD)
3) Is there a formal software technical requirements review prior to software development?
4) Is software reviewed in terms of top down and bottom-up?
5) Is quality designed into software?
6) Is there a formal process for debugging software?
7) Are the following tested in software?
   a) Ease of use
   b) Failure rate
   c) Fault density
8) Does a software quality management program exist?
9) Does a software quality verification of procured software and hardware exist?
10) Are computers and electronic data interchange used?
11) Is there a formal, documented quality program in the following:
   a) Computer equipment acquisition?
   b) Software?
   c) Data storage?
   d) Training?
   e) Documentation?
12) Does an interdisciplinary team deal with software and hardware purchasing and development?
13) Are thorough specifications and documentation developed?
14) Is there a recruitment and training program for computer personnel?
15) Are plans developed for:
   a) Equipment integration?
   b) Operating procedures?
   c) Programming and testing?
   d) Personnel training and development?
16) Is there a procedure for hardware maintenance?

17) Is information exchanged through computer networks?

18) Are computer data backed up?

19) Are computer personnel qualified and certified?

20) Do personnel know how to use computer equipment, including:
    a) Software?
    b) Hardware?
    c) Firmware?

21) Are computer systems compatible?

22) Is software compatible?

23) Is information backed up?

24) Is equipment periodically maintained?

25) Is equipment placed in proper locations for ease of use?

INSPECTION AND TESTING OF SUPPLIED PRODUCTS

1) Are incoming shipments sampled, inspected, and tested?
2) Are samples representative of the shipment?
3) Are samples pulled randomly or systematically?
4) If a shipment consists of discrete units, is a sample taken from two or more sample units?
5) If shipments are delivered in a continuous process, are samples pulled systematically throughout the shipments so that within-lot variation is included?
6) If the shipment is a single unit of material, are samples taken from two or more locations in the shipment?
7) Does special testing require approval from the customer engineering
8) department?
9) Is inspection and testing performed in accordance with control plans?
10) Are the following tests conducted?
    a) Physical
    b) Chemical
    c) Metallurgical
    d) Mechanical
    e) Dimensional
    f) Functional
    g) Assembly
    h) Regulatory
    i) Compliance
11) Are sampling, inspection, and testing frequency determined statistically?
12) Do inspection personnel understand:
    a) Engineering print reading?
    b) GDT interpretation (geometric dimensional tolerancing)?
    c) Measuring and test equipment use?
    d) Sampling tables?
    e) Measurement and test equipment?
13) Are shipping and receiving areas well lighted, clean, and safe?
14) Have incoming material acceptance procedures been developed?
15) Is there a procedure for altering inspection levels?

16) Do procedures account for supplier's process capability?
17) Do suppliers attach certifications or other statistical evidence with incoming shipments?
18) Has the customer developed formal lot acceptance policies with the supplier?
19) Are AQLs specified for different types of material?
20) Do suppliers maintain production, test, or inspection stamps?
21) Do suppliers use SPC?
22) Is capability index specified in a P.O. or other documentation?
23) Do suppliers provide SPC charts for critical and major quality attributes?
24) Does documentation define acceptance/ rejection criteria for incoming raw material?
25) Are rejected shipments clearly identified and effectively segregated?
26) If problems occur, are they resolved quickly and efficiently?
27) Are tests performed internally?
28) If an outside contractor conduct tests is the contractor qualified?
29) Are data retained over the useful lot of a part?
30) Are there regulatory requirements as to how long documents should be retained?
31) If yes, are documents retained over the required period?

INSPECTION PROCEDURES

1) Are the following evaluated prior to inspection?
   a) Instructions
   b) Materials
   c) Setup
   d) Sampling plan
   e) Sample size
   f) Selection of samples
   g) Calibration status
   h) Workmanship standards
   i) Inspector qualifications

2) Do inspection instructions address:
   a) Accept/reject criteria?
   b) Inspection and testing equipment?
   c) Drawing numbers and revisions?
   d) Detailed inspection instructions?
   e) Product characteristics to be inspected?
   f) Location of characteristics on print?
   g) Tolerance limits of quality characteristic?
   h) Environmental conditions of inspection?
   i) Precautions and safety practices?
   j) Inspection levels?
   k) Acceptable quality level (AQL)?

3) Do inspection records address:
   a) Inspected part name and part number?
   b) Drawings, number, and revision?
   c) Lot identification?
   d) Inspection date?
   e) Inspection and test conducted?
   f) Lot size?
   g) Sample size?
   h) Number of parts rejected?
   i) Reason for rejection?
   j) Inspection date?
   k) P. 0. Number?
   l) Quality characteristics inspected?

4) Are inspection/test records kept on file?
5) Are engineering prints readily available?
6) Is manufacturing documentation readily available?
7) Is sampling and inspection planned and conducted in the following areas:
   a) Incoming material?
   b) In process?
   c) Final?
   d) Laboratory?
   e) Machine setup?
   f) Packaging?
   g) Installed products?
   h) Service operations?
8) Is a written procedure developed for sampling?
9) If yes, does the procedure:
   a) Describe the sampling plan?
   b) List measurement equipment?
   c) List container requirements?
   d) Describe how to obtain a sample?
   e) State the required sample sizes?
   f) List measurements, analyses to be performed?
10) Is the sampling environment, for example the loading dock, safe?
11) Is the material being sampled toxic or hazardous?
12) Are inspectors protected from exposure to contamination?
13) Are inspectors aware of the potential contamination hazards of the sampled materials?
14) In terms of sampling, is the following provided or specified:
   a) Sampling tables?
   b) Sampling frequency?
   c) Sample sizes?
   d) Acceptable quality levels?
   e) Operating characteristic curves?
   f) Alpha and beta risks?
15) Is inspection static or patrolling?
16) Are visual aids used for inspection comparison purposes?
   a) Photographs
   b) Samples
   c) Exploded view drawings

    d) Three-dimensional figures

17) Are inspection levels monitored?

18) Is there statistical control documentation on all critical and major control characteristics?

19) Does inspection involve the following quality requirements?
    a) Physical
    b) Dimensional
    c) Understandable
    d) Accurate
    e) Complete
    f) Metallurgical
    g) Mechanical
    h) Chemical
    i) Visual
    j) Functional
    k) Fit

20) Do inspection instructions incorporate:
    a) Part number?
    b) Part name?
    c) Operation?
    d) Operator/inspector name?
    e) Quality attributes evaluated?
    f) Sample size?
    g) Inspection frequency?
    h) Acceptance criteria?
    i) Measurement/test equipment to use?
    j) Method to use?
    k) Type of deficiencies found?
    l) Standard for acceptance/rejection?
    m) Approved material disposition?
    n) Rejected material segregation?
    o) Notification date?
    p) Requirements?
    q) Inspection levels?
    r) Accept/reject criteria?
    s) Measuring equipment for each inspection?
    t) Inspection location?
    u) Special inspection operations?

    v) Data collection requirements?

21) Are sampling and inspection instructions:

    21.1) Does inspection follow-up incorporate the following?

    a) Inspection of reworked material

    b) Corrective action request

    c) Type and date of resolution

    d) Post audit requirement

22) Are instructions established detailing how reworked material is to be inspected and to what quality level?

23) Are quality studies conducted following rejections?

24) Does documentation verify the accuracy of the studies?

25) Is corrective action implemented for all out of control processes and conditions?

26) Are inspectors trained in:

    a) Sampling plans?

    b) Measurement?

    c) Safety?

    d) Analysis?

    e) Sampling techniques?

27) Are inspectors periodically certified?

28) Are the most current test/inspection instructions available to inspectors?

29) Do roving inspectors work with line personnel to solve problems?

LEGAL AND REGULATORY COMPLIANCE

1) Is there a process to identify legal, code, regulatory and other requirements that apply to programmatic, project, and activity level quality?

2) Is the Quality Manager responsible to ensure and demonstrate the appropriate level of compliance to health, safety, environmental, ethical, and security regulations?

LESSONS LEARNED POLICY

1) Has the organization established organizational or a programmatic process that describes the use of quality policies, objectives, internal audit results, analysis of data, corrective and preventive actions, and management review to facilitate continual improvement?

2) Is historical quality information and data secured and available in order to improve project processes?

3) Has the Quality Manager developed processes to ensure that key design, production, and quality activities satisfy stakeholder expectations?

4) Are variances, nonconformances, risks, other unusual data collected, analyzed, and communicated to appropriate stakeholders weekly?

5) Has the Quality Manager authorized the measurement and monitoring of project processes to determine:
   a) Stakeholder satisfaction?
   b) Risk mitigation?
   c) Cost containment?
   d) Project timeliness?
   e) Constructability?
   f) Safety?
   g) Environmental compliance?
   h) Legal compliance?

6) Does the Quality Manager periodically or as required authorize:
   a) Quality management reviews (QMRs)?
   b) Quality assurance reviews (QARs)?
   c) Quality control reviews (QCRs)?
   d) Quality Management reviews?
   e) Quality assurance reviews?
   f) Quality control reviews?

7) Can the Quality Manager specify or request contractors and/or personnel to identify areas of potential risk where quality assurance audits and control inspection, tests, and/or measurements be conducted?

8) Does the Quality Manager determine the value and authorize the specific inspection?

9) Are measurements recorded and reported to the appropriate parties if action is required?

10) Are quality assessments conducted against objective quality standards, specifications, policies, procedures, and work instructions?

MATERIAL DISPOSITION PROCEDURES

1) Is there a procedure for disposing of nonconforming material, including:
   a) Scrap?
   b) Rework?
   c) Salvage?
   d) Return to supplier?
   e) Reclaim?
2) Are reworked/scrap products audited?
3) Are waivers/deviations being eliminated?
4) Is nonconforming material segregated and tagged?
5) Are nonconforming products removed from the production line?
6) Is material identified as it goes through each process step?
7) If material is removed from a process, does documentation reveal where, how many, and why it was removed?
8) Are the following considered when dealing with nonconforming material:
   a) Proper segregation?
   b) Clear identification?
   c) Traceability to inspection report?
9) Do shipping records incorporate the following information:
   a) Part number?
   b) Part name?
   c) Manufacturer name?
   d) Manufacturer location?
   e) Purchase order?
   f) Quantity?
   g) Shipment location?
10) Does formal traceability program exist for:
    a) Raw material?
    b) Defective products?
    c) Failed products?
    d) Incoming material?
    e) In process material?
    f) Final material?
11) Is authority to reject nonconforming products defined and documented?
12) Are the following considered when material is rejected?

a) Rejection tag placed on shipment
b) Proper segregation
c) Inspection report completed

13) Are finished products tested prior to loading for shipment?
14) Are finished products tested after loading and prior to shipment?
15) Are certificates of analysis forwarded to the customer?
16) Is there a procedure to notify customer if nonconforming products are shipped?

MATERIAL DISTRIBUTION

1) Is there a formal procedure to evaluate the effectiveness of distribution challenges and methods?
2) Is there a push or pull demand system?
3) Are 'push' systems accurate in terms of forecasting demand?
4) Are 'pull' systems accurate in terms of ensuring smooth flow of products?
5) Are there formal supply management processes and procedures?
6) Are recyclable products sourced into new products?
7) Can raw materials be reused or recycled?
8) How much energy is required for this process?
9) Is transportation planned to reduce waste of resources?
10) Are transportation methods chosen to reduce environmental impacts and costs?
11) Are component parts identified for all manufactured products?
12) Can components be recycled?

MATERIAL LABELING

1) Are labels checked to ensure they satisfy customer, internal, and regulatory requirements?
2) Are guidelines established for proper labeling, including:
    a) Legal?
    b) Safety?
    c) Environmental?
    d) Health?
    e) Regulatory?
3) Is the following information provided on container labels?
    a) Contents
    b) Manufacturing date
    c) Origin
    d) Quantity
    e) Safety/hazard information
    f) Net weight
    g) Lot number
    h) Customer name and address
    i) Shelf life
    j) Proper handling, storage, and shipping
    k) Part number
    l) Part name
    m) Destination
    n) Quantity
    o) Supplier identification
    p) Pack date
    q) Engineering change level
4) Do shipping containers identify part numbers?
5) Does the packing slip show the quantity of samples shipped and the required initial sample recognition?
6) Numbers?
7) Can material packaged in large containers be identified through bills of lading or accompanying test certificates?
8) Is a labeling control system established for different types of transport?
9) Is bar coding used extensively?

10) Are stored products properly labeled so that they are traceable to the source?

11) Are samples properly labeled?

MATERIAL MOVEMENT AND HANDLING SYSTEM

1) Do suppliers have sufficient lead-time to deliver quality products?
2) Do suppliers have a production scheduling group with regard to:
   a) Raw materials?
   b) Work in process?
   c) Finished materials?
3) Does the group have authority and responsibility for coordinating product schedules on a daily or weekly basis?
4) Are production schedules driven by customer orders?
5) Do suppliers minimize daily changes to the production schedule?
6) Are supplier's production and delivery schedules matched to customer requirements?
7) Do suppliers deliver materials on a just-in-time (JIT) basis?
8) Are JIT shipping schedules driven on a daily basis?
9) Do suppliers reconcile their schedules and records with those of the customer?
10) Has responsibility for shipment of finished goods been assigned?
11) Do production scheduling, manufacturing, purchasing, and shipping functions communicate regularly to ensure material availability?
12) Are shipping plans upgraded to ensure JIT delivery?
13) Are the following computerized?
   a) Incoming supplier data
   b) Shipping schedules
   c) Notifications
14) Is the inventory system computerized?
15) Is bar coding used extensively?
16) Does the inventory system report daily on following inventories?
   a) Raw materials
   b) Work in process
   c) Finished goods
17) Is a physical inventory check per formed quarterly or yearly?
18) Is there a periodic check of inventory against the most current engineering change order or prints?
19) Are inventory levels appropriate to the production cycles?
20) Is there a formal program to:
   a) Increase inventory turnovers?

    b) Decrease inventory levels?

21) Are accurate quarterly forecasts sent to suppliers?

22) Are customer requirements reviewed periodically?

23) Are updated forecasts sent to suppliers?

24) Are purchase and blanket orders accurate, timely, and comprehensive?

25) Do suppliers have specific quantities and delivery dates at least a month in advance of delivery?

26) Are suppliers' deliveries monitored?

27) Do regulatory standards require lot traceability of critical parts?

28) Is this traceability directly connected to required tests and inspections confirming compliance to specifications?

29) Are all shipments properly identified?

30) Are products properly identified?

31) Are products identified and tracked through:
    a) Permanent marks?
    b) Route cards?
    c) Tags?

32) Are critical parts or products identified?

33) Do these parts require lot traceability?

34) Are critical parts identified in terms of groups of parts by manufacturing (day, month, and year)?

35) Is there a procedure to investigate, report, and eliminate premature or late deliveries?

MATERIAL PACKAGING

1) Is product packaging carefully designed?
2) Is product packaging appropriate for the market, production, and customer?
3) Is standardized packaging used to make optimal use of transport and storage space?
4) Do suppliers use returnable, recyclable, and degradable packaging?
5) Are packaging materials chosen for lower energy use?
6) Is packaging designed to be reusable?
7) Is paper and cardboard collected for recycling?
8) Are plastics, Styrofoam minimized in preference for popcorn, and other recyclable materials?
9) Is packaging checked to ensure it satisfies customer, internal, and regulatory requirements?
10) Are guidelines established for proper packaging, including:
    a) Legal?
    b) Safety?
    c) Environmental?
    d) Health?
    e) Regulatory?
11) Are packaging and packing materials designed to accommodate the product?
12) Is there a procedure for specifying proper packaging?
13) Do suppliers notify the customer of changes in packaging methods?
14) Do suppliers consistently satisfy customer's packaging requirements?
15) Has the customer approved suppliers' packaging specifications?
16) Do suppliers package materials appropriately by part number?
17) Does the customer use returnable containers?
18) Do suppliers use returnable containers?
19) Is there a procedure for ensuring that containers are clean and undamaged?
20) Are samples pulled from open containers that are properly segregated?
21) Are packages requiring inspection and testing identified?
22) Is there a procedure for preventing transportation and handling damage?

23) Have problems been reported in handling, storage, and shipping products?

24) Are there special handling equipment and storage facilities?

MATERIAL QUALITY

1) Does the Quality Manager authorize the development of quality processes to ensure that handling, storage, preservation and delivery of products to the production site and other facilities conform to product requirements and specifications?

2) Are suppliers of these products specifying special handling, storage, preservation and delivery requirements prior to delivery to the job site?

3) Are suppliers responsible for maintaining the integrity of supplied materials?

4) Does sign-off by agents or contractors absolve the supplier for providing conforming materials or services?

5) Can products released to production until all specified quality activities have been satisfactorily completed and related documentation is available and authorized?

6) Are quality activities for incoming inspection, handling, storage, preservation, and delivery specified in quality inspection procedures?

7) Do handling methods specify the following:
   a) Unique identification of product?
   b) Proper handling and transportation instructions?
   c) Protective covers and/or coatings?

8) Do packaging methods specify the following:
   a) Handling, transporting, and storage methods?
   b) Loading and unloading equipment?
   c) Environment conditions?
   d) Contents of materials?

9) Do storage methods specify the following:
   a) Proper storage areas?
   b) Authorizing receipt to and from storage areas?
   c) Storage considerations?

10) Does the Quality Manager review incoming material specifications for production, maintenance, operations, and other activities to ensure that potential risks and nonconformances have been identified?

11) Does critical equipment have special handling, packaging, storage, preservation, or delivery instructions listed in the purchase order and incoming documentation?

MATERIAL SHIPPING PROCEDURES

1) Do suppliers operate a shipping and storage control program?
2) Are shipping instructions clear?
3) Are outgoing and incoming shipments coordinated and synchronized with internal operations?
4) Are the most effective, efficient, and economical methods for moving materials analyzed?
5) When shipping materials, are the following requirements considered and analyzed:
   a) Refrigeration?
   b) Damage-free freight cars?
   c) Piggyback truck-railcars?
   d) Containers?
6) Has the supplier considered the pack aging, shipping, labeling, and handling of the following?
   a) Flammable materials
   b) Corrosive materials
   c) Toxic materials
   d) Other hazardous materials
7) Is there an active and documented program of Department of Transportation (DOT) compliance?
8) Do supplier storeroom personnel have access to customer-specified packaging instructions?
9) Can suppliers implement a computerized system of shipment authorization?
10) Are packages periodically and randomly counted?
11) Do suppliers ship to the correct customer loading dock?
12) Do the supplier and carrier have to de liver material in a customer-specified sequence?
13) Are there sufficient personnel and receiving/shipping docks for the order levels?
14) Are excessive or premium freight charges investigated, reported, and eliminated?
15) Are transportation and delivery monitored and performance reviewed?
16) Can supplied materials be tracked to final products and customers?
17) Do batches or product shipments contain commingled material?

18) Are nonconforming materials segregated from conforming materials?

19) Are supplied items sequentially identified?

20) Are field or customer rework records maintained?

MATERIAL STORAGE PROCEDURES

1) Do handling, storage, and movement procedures address material risk?
2) Are products safe from humidity, temperature, particulates, and other physical conditions?
3) Does the customer consider how the product will be:
   a) Transported?
   b) Displayed?
   c) Stored?
   d) Secured?
   e) Labeled?
   f) Packaged?
4) Do suppliers consider how the product will be:
   a) Transported?
   b) Displayed?
   c) Stored?
   d) Secured?
   e) Labeled?
   f) Packaged?
5) Are procedures developed and personnel trained in material?
   a) Use
   b) Movement
   c) Testing
   d) Sampling
   e) Disposition
   f) Labeling
   g) Packaging
   h) Loading/unloading
6) Does the product have a specified shelf life?
7) Is information gathered or tests con ducted to determine shelf life?
8) Do products go beyond the stated shelf life?
9) Is shelf life determined by:
   a) Regulatory mandates?
   b) Statistical analysis?
   c) Customer surveys?
   d) Are products recalled quickly?

e) If products are for industrial use and are containers clean?

f) Are products hazardous or toxic?

10) If hazardous products are stored, are personnel instructed in:

   a) Transport?

   b) Handling?

   c) Storage?

MATERIAL TRANSPORTATION

1) Are anti-pollution energy saving measures implemented throughout the organization?
2) Are alternative fuels, such as natural gas or electricity used in transporting equipment?
3) Is there a preventive maintenance program for all types of equipment?
4) Is transportation of sourced products planned?
5) Are the movement of goods and components internally optimized for cost, delivery, and performance?
6) Is the movement of goods and components externally optimized for cost, delivery, and performance?
7) Are there formal procedures for developing appropriate packaging?
8) Are environmentally acceptable materials used in packaging?
9) Are under packaging or over packaging considered in package design?
10) Are marketing and sales strategies sensitive to cultural, ethnic, environmental, and other considerations?

MEASUREMENT AND TEST EQUIPMENT

1) Is new measurement and test equipment reviewed prior to use?
2) Does production equipment have sufficient production capacity to handle anticipated production volumes?
3) Is the equipment process capable?
4) Is SPC measurement equipment identified?
5) Is test equipment certified?
6) Is test equipment specified in engineering specifications?
7) Are requirements established for special gages and fixtures?
8) Are gage methods compatible between suppliers and customer?
9) Are inspection gages keyed to engineering drawings and control plan?
10) Are measurement system variation studies planned?
11) Can measurement systems discriminate to one-tenth of the engineering tolerance or better?
12) Are gages available sufficiently early to conduct preliminary process capability studies at the equipment supplier's facility?
13) Is correlation of all test equipment established?
14) Do calibration records address:
    a) Equipment identification?
    b) Engineering change number?
    c) Frequency of check?
    d) Equipment tags and labels?
    e) Adjustment when out of tolerance?
    f) Corrective action report?
    g) Calibration date?
    h) Calibrator name?
    i) Date of next calibration?
15) Are the following calibrated:
    a) Electronic test equipment?
    b) Tools and gages?
    c) Personal measuring equipment?
16) Is all measurement and test equipment certified?
17) Are gage certification documents available?
18) Are gage capability studies conducted?
19) Is measurement and test equipment approved by customer's engineering department?

20) Are routine calibration frequencies established?
21) Is machine capacity and potential established at the supplier's manufacturing facility before delivery?

MEASUREMENT AND TEST PROCEDURES

1) Have standard procedures been established for:
   a) Measurement?
   b) Laboratory tests?
   c) Mechanical and physical tests?
   d) Inspection?
2) Within the laboratory, test or inspection area, are these procedures consistent among all employees?
3) If there are multiple testing stations, laboratories or inspection areas, are measurement, and testing procedures consistent among all areas?
4) Are analytical, measurement, test methods, and procedures thoroughly documented?
5) Are the following records kept?
   a) Test
   b) Measurement
   c) Laboratory
   d) Calibration
   e) Traceability
   f) Repeatability
   g) Accuracy
   h) Precision
6) Are inspection or test records:
   a) Available?
   b) Current?
   c) Detailed?
   d) Understandable?
   e) Accurate?
   f) Comprehensive?
   g) Applicable?
   h) Used?
7) Is equipment precise and accurate?
8) Are repeatability studies periodically performed?
9) Do testing, inspection, laboratory personnel use approved methods to ensure accuracy, and precision?
10) Is test and measurement equipment validated before each use?
11) Are test methods verified against standards before each use?

12) Are nonstandard testing or measurement methods used?

13) Has management or the appropriate regulatory authority reviewed and approved the methods?

14) Has a regulatory authority issued a test variance?

15) Does present testing comply with the variance?

16) Has nonstandard testing become the norm?

17) If yes, is the modified testing approved to become the new testing method?

18) Has measuring equipment been checked for accuracy and precision?

19) If fixtures or patterns are used as part of a measurement device, have they been qualified?

20) Is there a procedure for determining the traceability of measuring equipment?

21) Has the auditee designated a primary standard for specified components, materials, testing equipment, and so forth?

22) Has a hierarchy of standards been developed?

23) Is test and measurement equipment calibrated every:
    a) Six months?
    b) One year?
    c) Two years?

24) Do records indicate calibration history?
    a) Are the following reference standards traceable to a third-party organization: Reference standards?
    b) Reference measurement equipment?

25) Is calibration traceable to NIST?

26) Is the calibration interval sufficient to ensure measurement instrument accuracy and precision?

27) Are industry prescribed testing and measurement techniques used?

28) Is the following calculated on test and measurement equipment?
    a) Repeatability
    b) Reproducibility

29) Is test and measurement equipment properly:
    a) Secured?
    b) Stored?
    c) Tagged?

30) If variation is excessive following equipment evaluation, is corrective action initiated?

31) Did corrective action eliminate recurrence of the problem?

MEASUREMENT EQUIPMENT CONTROL

1) Do company and contractor production measurement equipment comply with the following:
   a) Critical quality measurement areas are located in print, quality plan, etc.?
   b) Critical measurement equipment is identified?
   c) Measurement equipment is calibrated and adjusted at specified intervals?
   d) Method of calibration is specified?
   e) Calibration results are specified?
   f) Proper environmental measurement conditions are considered when conducting measurements?
2) Critical measurement are revalidated if required?

NONCONFORMANCE CONTROL

1) Does the Quality Manager designate quality personnel to evaluate each of the following?
2) Nonconformance to standards, specifications codes, policies, standards, and procedures:
    a) Variances in quality?
    b) Quality trending?
    c) Risk levels?
    d) Waiver and concessions?
    e) Supplied product nonconformance?
    f) Corrective action requests?
    g) Prevention action requests?
    h) Audits results?
3) Are quality responsibilities and authorities for review and resolution of nonconformances defined?
4) Do quality personnel review nonconformances and determine actions to be taken, including:
    a) Correct to conform to code or specifications?
    b) Accept under waiver or concession?
    c) Repair/rework?
    d) Accept with or without correction (use as is)?
    e) Reassess for further review?
    f) Reject as unsuitable?
    g) Record determination?
5) Do nonconformance/risk assessment:
    a) Identify, record, and review the nonconformance?
    b) Analyze the risk and prioritize each nonconformance?
    c) Review repetition of occurrences?
    d) Conduct trend analysis?
    e) Analyze the root cause?
    f) Determine the next steps to prevent recurrences?
    g) Compile lessons learned?

PREVENTIVE ACTION REQUESTS

1) Has the Quality Manager established a process for tracking and preventing design, production, operational, and supplier nonconformances?
2) Are preventive action requests required of nonconformances resulting from:
   a) Stakeholder complaints?
   b) Nonconformance reports?
   c) Quality Management System review results?
   d) Quality assurance review reports?
   e) Quality control review reports?
   f) Data analysis results?
3) Do preventive action requests.
4) Identify potential nonconformances:
   a) Identify designs and/or production risks
   b) Determine causes of potential nonconformances and/or risks
   c) Authorize investigation and recommend risk mitigation practices
   d) Determine preventive actions needed to eliminate causes of potential nonconformances
   e) Implement preventive action
   f) Record, verify, and validate results

PROJECT CLOSEOUT

1) Does the Quality Manager work with stakeholders to communicate and coordinate the smooth transfer of constructed facilities and equipment?
2) Does senior management develop a quality plan to ensure a smooth transfer of the following:
    a) Facilities?
    b) Special equipment?
    c) Performance criteria?
    d) Security requirements?
    e) Process requirements/
    f) People requirements?
    g) Operating/maintenance requirements?
3) Is special equipment identified, tagged, and traceable to engineering prints and operating instructions?
4) Do operational stakeholders have sufficient information to operate constructed facilities and establish a preventive maintenance program?
5) Upon stakeholders request, do contractors may be requested to provide preventive maintenance procedures and operate equipment until handoff?
6) Does the Quality Manager work with stakeholders to ensure a smooth hand off of all design, production, operations, and maintenance information?
7) Does the Quality Manager have the authority to initiate changes to maintain quality consistency?
8) Can critical design or production changes be verified to ensure that the instituted change had the desired effect?
9) Are there critical gates of the design/production/closure process?
10) Does the Quality Manager verify the following:
    a) Smooth transfer of clear and understandable working or installation instructions?
    b) Complete and accurate design and production specifications?
    c) Maintenance, installation, and service equipment?
    d) Suitable measurement or monitoring equipment?
    e) Acceptable methods of release and delivery and/or installation of product/service?

PRODUCTION AND MANUFACTURING

1) Is there a plan to acquire the necessary production and manufacturing resources?
2) Are suppliers selected, monitored, and improved throughout the production lifecycle?
3) Have the following suppliers been chosen?
   a) Capital goods
   b) MRP
   c) Materials
   d) Services
   e) Packaging
   f) Warehousing
   g) Logistics
4) Have 'make or buy' decisions been made for critical components?
5) Have in-house or supplier assemble decisions been made?
6) Has the appropriate technology been acquired to manufacture the product?
7) Is there a plan to arrange product shipments?
8) Is there're a contingency plan for:
   a) Alternate suppliers
   b) Alternate distribution
   c) Alternate warehousing
9) Are there sufficient people and other resources to deploy the plan?
10) Is the following planned, and monitored?
    a) Inventories
    b) Packaging
    c) Maintenance
    d) Installation
    e) Operations
11) Does the floor plan indicate all required process and quality control points?
12) Do process and quality control stations include all necessary equipment and files?
13) Are there adequate staging and impound areas?
14) Are quality control points located to prevent shipment of nonconforming products?

15) Is there a potential for nonconforming products to commingle with conforming products?
16) Can overhead material or air-handling systems contaminate exposed product?
17) Are there final audit facilities?
18) Is there a process flow diagram that illustrates the sequence of production, inspection, transportation, storage, and inspection operations?
19) Is process failure modes effects analysis (FMEA) used to design process flow?
20) Is the process flow diagram in the control plan keyed to the process and product evaluation points?
21) Is material movement clearly identified on the flow diagram?
22) Are there provisions to identify and inspect reworked products before use?
23) Are statistical control points identified on the process flow diagram?
24) Are material handling, labeling, storage, movement, and other problems identified and corrected?

PRODUCTION DOCUMENTATION

1) Do manufacturing personnel understand specifications?
2) Can manufacturing personnel identify critical and major parameters in engineering drawings?
3) Are instructions developed for:
   a) Setup?
   b) Incoming material inspection?
   c) In-process control?
   d) In-process inspection?
   e) Layout?
   f) Final/outgoing quality?
   g) Nonconforming product disposition?
4) Is the following manufacturing documentation available?
   a) Work instructions
   b) Inspection instructions
   c) First article inspection
   d) Process control charts
   e) Preventive maintenance records
   f) Tooling records
   g) Packaging instructions
5) Does the work packet traveling with processed material include:
   a) Sequence of operations?
   b) Inspection and test instructions?
   c) Inspection points?
   d) Setup, equipment, and tooling requirements?
   e) Process speeds and feeds?
   f) Drawing numbers and revision?
   g) Updated engineering prints?
   h) Quality standards and specifications?
   i) Raw fabricated, or assembled material specifications?
6) Are quality characteristics identified on engineering drawings?
7) Are quality levels specified?
8) Is there a list of procedures to be completed for special inspection, testing, maintenance, and shutdown?
9) Are operators consulted in developing instructions?
10) Are work process standards and instructions developed?

11) Do process control instructions exist at all workstations?

12) Are process control instructions current and complete?

13) Is each job documented?

14) Are work standards and instructions written?

15) Is responsibility to approve process changes clearly defined and followed?

16) Is there a procedure to notify the customer with advance notice of process changes?

17) Does document and drawing control address:

    a) Drawing number and revisions?

    b) Procedures for control, storage, revision, numbering, and revision?

    c) Disposal and use of obsolete drawings?

PROCESS MONITORING, CONTROL, AND IMPROVEMENT

1) Is regulatory approval necessary for the quality program, processes, or product?
2) Are significant quality characteristics that affect customer-perceived fit and function identified?
3) Are production process characteristics that affect product characteristics identified?
4) Are these quality characteristics keyed to the control plan and to the process flow diagram?
5) Are specifications reviewed when:
   a) Process performance improves?
   b) Process capability decreases?
   c) Customer needs change?
   d) Problems recur?
   e) Corrective action fails?
6) Does the auditee have a process flow diagram?
7) Does the process flow diagram indicate where quality is monitored?
8) Does the process flow diagram indicate how quality is controlled?
9) Does the process flow diagram indicate who controls quality?
10) Are quality characteristics prioritized on prints and technical documents?
11) Are the following statistical methods used in problem solving?
   a) Pareto analysis
   b) Histograms
   c) Cause and effect analysis
   d) Regression analysis
   e) Design of experiments
12) Is SPC used for process control?
13) Are all critical processes:
   a) Under statistical control?
   b) Capable?
   c) Improved?
14) Are critical process parameters identified?
15) Are they charted?
16) Are process capability goals stated?

17) Are the following readily available or posted:
   a) Specification standards?
   b) SPC charts?
   c) Critical parameters?
18) Are SPC charts maintained for critical parameters?
19) Are they attached to shipments?
20) Are processes reviewed for control and capability?
21) Are manual processes being automated?
22) Are production processes fully documented?
23) Are service processes fully documented?
24) Do people know their internal customers?
25) Is there a systematic effort to satisfy internal customers?
26) Is there a systematic training program?
27) Do process monitoring and control documents address the following:
   a) Operator training?
   b) Operator instructions?
   c) Product instructions?
   d) Sequence of operations?
   e) Material identification?
   f) Material segregation?
   g) Safety precautions?
   h) Product inspection?
   i) Machine maintenance?
   j) Quality requirements?
   k) Quantity requirements?
   l) Measurement and gage instructions?
28) Are the following process monitoring and control instructions used?
   a) Engineering drawings
   b) Process control sheets
   c) Inspection instructions
   d) Test inspections
   e) Production traveler
   f) Control plan
29) Is machine setup controlled?
30) Are documents understandable?
31) Are operators certified?
32) Are there internal control systems separated from statistical process control systems?

33) Are these systems closed loop?

34) Do they provide real-time feed-forward or feedback information?

35) Is this information used to control and improve the systems?

36) Has process capability been defined?

37) Are instructions and procedures written?

38) Is there a safety checklist for critical process steps?

39) Is there a list of raw materials or components for each process step?

40) Is equipment periodically inspected and maintained?

41) Has everyone been trained in SPC?

42) Are the following charts used to monitor quality?
   a) X-bar and R
   b) Cumulative sum (CUSUM)
   c) Standard deviation
   d) Moving range (MR)
   e) Moving average (MA)
   f) Exponentially Weighted Moving
   g) (EWMA)
   h) Multiple variable

43) Are charts manually charted?

44) Are charts computerized?

45) Is corrective action initiated to remove chronic problems?

46) Are control charts periodically reviewed for performance?

47) Are the specifications and tolerances periodically reviewed?

48) Does engineering use process capability to develop specifications?

49) Are control limits modified as the process improves?

50) Are charts with hourly sampling reviewed:
   a) Weekly?
   b) Monthly?
   c) Bimonthly?

51) Are charts with daily sampling reviewed:
   a) Weekly?
   b) Monthly?
   c) Quarterly?
   d) Yearly?

52) Is the sample size periodically reviewed?

53) Does SPC sampling frequency detect changes?

54) Are all processes in statistical control?

55) Have capability indices been established for critical quality characteristics?

56) Is continuous improvement pursued?

57) Is process capability used to allocate internal resources?

58) Is process capability used to determine if specifications can be met?

PRODUCT RELIABILITY TESTING

1) Is there a formal program to improve reliability of existing and new products?
2) If yes, are the following tests conducted:
   a) Accelerated life?
   b) Failure mode and effects analysis?
   c) Derating?
   d) Redundancy?
   e) Fault tree analysis?
3) Is safety engineering and analysis conducted on all major process and products?
4) Are the following regulations monitored and adhered to:
   a) OSHA?
   b) FDA?
5) Are mechanical safety issues addressed, including:
   a) Safe handling?
   b) Machine guards?
   c) Protective clothing?
   d) Breathing equipment?
   e) Explosion-proof cabinets and fixtures?
6) Is reliability measured in terms of:
   a) Mean time before failure (MTBF)?
   b) Failure rate?
   c) Mean time to first failure?
7) Are product loads and stresses considered during:
   a) Startup?
   b) Normal operation?
   c) Emergency operation?
   d) Shutdown?
   e) Shipping?
   f) Storage?
   g) Movement?
8) Are environmental conditions on product reliability considering during:
   a) Startup?
   b) Normal operation?
   c) Emergency operation?

d) Shutdown?

e) Shipping?

f) Storage?

g) Movement?

9) Are the most severe conditions anticipated in reliability design?

10) Is a design margin calculated for each severe condition?

11) Are design characteristics that affect high-risk priority failure modes identified?

12) Has a design wish list been prepared?

13) Do the effects of the failure modes address all the elements on the wish list?

14) Are the most likely failure modes identified, including:

a) Failure to operate at prescribed time?

b) Failure to cease functioning at prescribed time?

c) Premature operation?

d) Operation in an incorrect manner?

15) Is design of experiments (DOE) analysis used to prioritize major causal factors?

16) Have contingencies been identified to reduce risk of each of the above failures?

17) Does design failure modes effects analysis (FMEA) assess design adequacy and avoid causes and controls that belong to the process?

18) Has product reliability been calculated?

19) Is reliability testing conducted at these levels?

a) Product

b) Systems

c) Assemblies

d) Subassemblies

e) Components

20) Are areas of greatest risk or costs identified?

21) Is there a formal effort to track and incorporate changes into drawings?

22) Are reliability patterns or trends analyzed?

PRODUCT TESTING

1) Are the following tests conducted:
    a) Acceptance?
    b) Stress?
    c) Prototype?
    d) Development?
    e) Scale model?
    f) Wireframe?
    g) Simulation?
2) Is a test plan developed for each major test?
3) Does the test plan: identify customer requirements?
    a) Specify the best methods for obtaining information
    b) Recommend methods for testing
    c) Define methods of measurement
    d) Select test facilities
    e) Identify test personnel
    f) Define test environment
    g) Identify risks
4) Is computer testing and simulation conducted?
5) Is testing conducted internally or externally?
6) Do external testing facilities comply with internal requirements?
7) In conducting tests, are the following considered?
    a) Personnel training
    b) Test and measurement equipment selection
    c) Traceability of test equipment
    d) Equipment calibration
    e) Accuracy and precision
    f) Recording data
    g) Safety
    h) Regulatory approvals
    i) Training
    j) Data collection
    k) Test records

PRODUCTION/DELIVERY OF SERVICES

1) Is the service lifecycle managed?
2) Is there a plan to acquire the necessary resources to deliver high quality services?
3) Are suppliers selected and registered to ISO 9000 or some other standard?
4) Is the appropriate technology used to deliver customer-satisfying services?
5) Are there sufficient people with at the right core competencies to deliver a high quality service?
6) Is there a plan to develop the necessary skills if there are some deficiencies?
7) Are training and people resource requirements aligned with the corporate direction and strategy?
8) Are employees responsible for their training and development?
9) Is there a plan on how each customer segment will be satisfied?
10) Is the quality of service measured and corrected as required?

PRODUCTION ENVIRONMENT

1) Is there an internal housekeeping program?
2) Is the appearance of the work environment conducive to quality work?
3) Is the work environment clean?
4) Are there regulatory requirements for special facilities?
5) Are the following sufficient:
   a) Space?
   b) Lighting?
   c) Noise control?
   d) Safety requirements?
6) Is the work area accessible to all personnel?
7) Are internal supplier and customer groups closed to each other?
8) Are customer service personnel located close to the service area?
9) Is there sufficient space for expansion or adjustment of work?
10) Are aisles wide enough for traffic and people movement?
11) Are fire exits marked and unobstructed?
12) Are restroom facilities clean and convenient?
13) Are restroom facilities available for customers and visitors?
14) Are there common area conference areas?
15) Do offices have sufficient cabinets, computer terminals, copiers or other work equipment?
16) Can offices be rearranged through the use of modular furniture?

PRODUCT AND WORKPLACE SAFETY

1) In communicating safety directions to employees, are the following considered:
   a) Educational level?
   b) Culture?
   c) Languages?
   d) Environment?
   e) Type of usage?
2) Are areas of high risk and exposure identified?
3) Are sanctions or litigation pending with:
   a) Companies?
   b) Suppliers?
4) Is there a process of risk control and damage containment?
5) Does the organization monitor federal, state, and municipal regulations concerning:
   a) Environment?
   b) Safety?
   c) Affirmative action?
   d) Health?
6) Does the organization monitor compliance?
7) Is there a liability prevention program?
8) Are there existing claims or judgments against the company?
9) Are liability costs tracked?
10) Are liability costs high in relation to similar companies?
11) Is there sufficient insurance to cover eventualities?
12) Does the organization have a safety manual?
13) Does the organization have a safety department?
14) Are safety rules established for all company areas?
15) Are special evacuation routes identified?
16) Are safety drills periodically conducted?
17) Is safety information provided to the company's customers?
18) Is corrective action effective and efficient?
19) Have there been product recalls?
20) Are the following analyzed for liability exposure?
    a) Product design
    b) Packaging

    c) Shipping

    d) Storage

    e) Installation

    f) Service

    g) Recalls

    h) Field failure

    i) Supplier material

    j) Packaging

    k) Shipping

    l) Customer complaints

    m) Installation

    n) Customer performance

21) Is there an audit trail for the above documents?

22) Are defective materials retrieved and tested as a result of:

    a) Safety concerns?

    b) Poor product performance?

    c) Regulatory action?

23) Is there a procedure for resolving problems quickly?

24) Are audits efficient, effective, and economic?

25) Has a product failure resulted in:

    a) Health violations?

    b) Safety violations?

    c) Litigation?

    d) High monetary damages?

26) Is there a companywide quality measurement program?

27) Are the following safety precautions used?

    a) Electrical interlocks

    b) Mechanical stops

    c) Safety instructions

    d) Bulletins

    e) Illustrations

    f) Labels

28) Are the following environmental conditions considered in test and measurement:

    a) Temperature?

    b) Humidity?

    c) Particulates?

    d) Vibration?

29) Do "clean rooms" meet industry and regulatory requirements?
30) Is there a formal, documented calibration program?
31) Are safety warnings properly placed on packages and labels?
32) Has the legal department evaluated the safety warnings?
33) Do the warnings comply with regulatory requirements?
34) Is there a preventive maintenance program?
35) Is there a program to replace worn parts?
36) Are experimental and field data used to predict failure rates?
37) Is field and customer information used to improve products?
38) Are failed field products analyzed to determine the real cause of the problem?
39) Are customer complaints handled expeditiously?
40) Is customer satisfaction measured?

PRODUCTION/MANUFACTURING

1) Are manufacturing, production, and sourcing processes flowcharted?
2) Are manufacturing, production, and sourcing processes analyzed for improvement and for lean production?
3) Are innovative practices, emerging technologies, new processes considered in product development?
4) Is the company complying with or exceeding existing environmental protection standards?
5) Are there procedures to reduce harmful emission in manufacturing, distribution, etc.?
6) Is energy efficiency reduced by insulating, sealing, or otherwise protecting equipment and materials used at high or low temperatures?
7) Are all leaks in existing equipment for circulating or storing gases and liquids checked and repaired?
8) Is pollutant treatment equipment regularly checked for proper operation and efficiency?
9) Is noise produced machinery or equipment isolated or insulated?
10) Is there a process for reducing equipment vibration?
11) Is there a formal safety program?
12) Are employees and contractors trained in first aid, workplace hazards, etc.?
13) Are first aid equipment and stations conveniently located?
14) Are employee health complaints monitored for early indication of hazardous or toxic contact?
15) Are blood and urine samples checked for toxic or other chemicals?
16) Is the product lifecycle managed?
17) Can the functional life of the product be enhanced?
18) Is incoming, in process and final inventory managed?
19) Are finished and in process products stored so wear and tear are minimized?
20) Are products designed for?
    a) Aesthetics
    b) Reliability
    c) Maintainability
    d) Durability
    e) Reparability

    f) Disposability

    g) Sustainability

    h) Ease of use

21) Are design, color, or other templates developed?

22) Is the product brand managed throughout the lifecycle?

23) Is there a plan for disposing of obsolete products?

PRODUCTION MATERIALS

1) Are hazardous or environmental damaging characteristics of materials identified?
2) Is the organization in compliance with Federal and state laws?
3) Is information collected of potential hazards of material used in production?
4) Are special or potential ecological problems identified from disposing into the atmosphere?
5) Are potential pollutants monitored?
6) Do suppliers provide material information?
7) Are improvement ideas solicited from suppliers?
8) Is information on alternative products, raw material substitutions, and other information collected?
9) Is there a procedure for the onsite storage of raw materials?
10) Are hazardous substances appropriate labeled and is access restricted?
11) Are maintenance and storage facilities designed for safety?
12) Is advance equipment installed in storage areas to ensure product integrity?
13) Are integrated efforts used to improve the efficiency and effectiveness of product use?

PRODUCTION PLANNING

1) Is there a preproduction quality planning procedure?
2) Are sufficient personnel dedicated to:
   a) Operate processes/machinery?
   b) Maintain processes/machinery?
   c) Implement control plan requirements?
   d) Inspect products?
   e) Test product performance?
   f) Analyze problems?
3) Are operators sufficiently trained to perform statistical process control, capability studies, and problem solving?
4) Is each operator and inspector provided instruction sheets keyed to the control plan?
5) Are sampling and test instructions placed near the operator?
6) Are the latest drawings, specifications, and other relevant information at the point of inspection?
7) Is there a procedure to maintain and establish reaction plans for SPC?
8) Is new equipment tested prior to introduction into production?
9) Are product substitutions approved prior to production?
10) Are products tested at or near specification limits?
11) Is testing conducted on:
    a) Final products?
    b) Assemblies?
    c) Subassemblies?
    d) Components?
12) Are tests conducted at:
    a) Conception?
    b) Feasibility?
    c) Prototype?
    d) Preproduction?
    e) Pilot?
    f) Field (beta site) test?
13) Are the following tests and/or evaluated prior to actual production?
    a) Operational method
    b) Inspection methods
    c) Machinery

    d) Dies

    e) Operators

    f) Measurement equipment

    g) Automated equipment

    h) Material movement equipment

    i) New measurement equipment

    j) Prototypes

    k) Modified products

14) Can products be manufactured with commercially proven, available equipment?

15) Can manufacturing or suppliers consistently produce products based on production volumes and schedules?

16) Is the proposed manufacturing process cost effective, efficient, and economical?

17) Can products be produced as specified on drawings with a minimum amount of explanation?

18) Is there a Cpk of 1.33 for all critical and major characteristics?

19) Can specified requirements be met at projected volume levels?

20) Are manufacturing processes analyzed to support production at required volume and quality levels?

21) Does design allow use of conventional material handling equipment and techniques?

22) Is there a preproduction material, handling, and storage procedure?

23) Is there a production testing procedure?

24) Are tests conducted by trained personnel?

25) Is instrumentation accurate and precise?

26) Has instrumentation been calibrated?

27) Is the test methodology complete?

28) Are test results used in quality improvement?

29) Are initial production shipments identified properly after initial sample approval?

PRODUCTION PREVENTIVE MAINTENANCE

1) Is a preventive maintenance program used on all major equipment?
2) Are preventive maintenance documents maintained?
3) Are machine or process shutdowns scheduled so equipment can be inspected?
4) Is there a history of unplanned process interruptions?
5) If yes, what was done to minimize or eliminate the interruptions?
6) Do operators self-manage and self-inspect their equipment?
7) Are tooling and fixtures properly stored and secured?
8) Are first products from a machine or process inspected?
9) Does a deficient first product trigger a tool maintenance program?
10) Is test equipment available?
11) Are tool/die records maintained, accurate, and complete?
12) Do chronic problems occur?
13) Is corrective action pursued when problems occur?
14) Does corrective action eliminate chronic problems?
15) Are results of corrective action evaluated?
16) Are parts available to make quick and effective repairs?
17) Are trained personnel and equipment available to make quick and effective?
18) Are downtime or shutdowns analyzed to discover root causes?

PRODUCTION QUALITY

1) General requirements:

a) Does senior management prepare production plan, control production, and production quality activities?

2) Does the Quality Manager prepare production quality documents, which at a minimum address the following:

a) Identify critical review gates of the production process?

b) List required production review, verification, and validation activities at each gate?

c) Identify responsibilities and authorities for production quality activities?

d) Coordinate interfaces among production managers, designers, stakeholders, etc.?

e) Clarify coordination, communication, and resolution responsibilities?

PRODUCTION QUALITY REVIEW

1) Are production quality requirements defined at key production points/gates on a quality plan, schedule chart or engineering print?
2) Are production quality assurance and inspection requirements reviewed for adequacy and completeness?
3) Are incomplete, ambiguous, or conflicting requirements resolved prior to proceeding to the next production step?
4) Are production requirements reviewed including:
   a) Production/code and or other requirements?
   b) Regulatory test/inspection requirements?
   c) Application of environmental requirements?
   d) Review of lessons learned from previous similar production activities?
   e) Ability to satisfy design requirements?
   f) Define areas of production risk and safety?
5) At defined gates of the production process, do systematic quality control reviews:
   a) Identify production quality requirements?
   b) Anticipate production problems?
   c) Identify production inspection/test locations?
   d) Plan and Identify solutions before actions are taken? .
   e) Record subsequent follow-up action?
   f) Ensure accurate red-lines?
   g) Control production changes?
   h) Review and authorize field changes to engineering prints?
   i) Control and communicate production changes?
6) Has production management and quality personnel designated trained personnel to conduct quality control checks, tests, and measurements?
7) Is production measurement and monitoring equipment controlled, calibrated, and properly maintained?
8) Is there a master list of critical measurement equipment with calibration dates maintained at the quality office?
9) Is measuring equipment accuracy and precision monitored on a regular basis?

PRODUCTION REVIEW AND CONTROL PLANS

1) Does process review and planning address:
   a) Process flowcharts?
   b) Design, process, or material feasibility?
   c) Failure mode and effects analysis
   d) (FMEA)?
   e) Control plans?
   f) Gauging, measurement, and test equipment quality plans?
   g) Preliminary process capability?
   h) Process monitoring and control instructions?
   i) Packaging/labeling/shipping/storage plans?
   j) Initial sample approval requirements?
   k) Prototype quality requirements?
   l) Incoming raw material and parts control?
   m) Purchased production product control?
   n) Ongoing quality planning?
2) For the entire process, is there:
   a) Sampling schedule and instructions?
   b) In-process statistical control?
   c) Identified critical characteristics?
   d) Actions to be taken based on process control conditions?
   e) Process records?
3) Are product quality characteristics identified?
4) Are procedures written and used?
5) Are final products inspected?
6) Are processes monitored?
7) If inspected, is the inspection based on a predetermined sampling plan and frequency?
8) If audited, is the audit based on in process statistical capabilities?
9) Does corrective action solve root cause problems?
10) Has a process flowchart been developed?
11) Have process instructions been developed for each step?
12) Are numerical targets with process parameters established for each critical step?
13) Has workflow been analyzed recently?
14) Are time studies used in analyzing workflows?

15) Are employees located near their work?

16) Is material located near work?

17) Have repetitive jobs been eliminated?

18) Has work been designed to reduce redundancy and unnecessary work?

19) Is workflow smooth?

20) If work is held up, is something being done to smooth it?

21) Is information leaving each process step accurate?

22) Does work go to the worker or does the worker have to chase it down?

23) Is good housekeeping practiced?

24) Is safety emphasized?

25) Is there a step-by-step description of the entire process?

26) Are machine and equipment capacity based on current sales?

27) Can manufacturing increase capacity without purchasing new plant and equipment:
   a) By 5%?
   b) By 10%?
   c) By 20%?

28) If volumes increase, does manufacturing have:
   a) Trained personnel?
   b) Sufficient raw material?
   c) Sufficient supplied material?
   d) Sufficient tooling?

PROJECT MANAGEMENT RESOURCES

1) Has senior management provided resources needed to establish and maintain the Quality Management System? Resources may include people, contractors, information, and financial resources?

2) Are trained, skilled and experienced personnel assigned to ensure development and deployment of the Quality Management System?

3) Does the Quality Manager:
   a) Evaluate legislation, regulations, standards, and policies that may affect organizational quality activities?
   b) Analyze programmatic and project level quality requirements?
   c) Determine and secure human as well as other resources?
   d) Determine quality competencies of existing organizational people to perform quality activities?
   e) Determine additional training needs?
   f) Provide quality training to address identified quality needs?
   g) Evaluate the effectiveness of quality training at defined intervals?
   h) Maintain appropriate records of quality training, skills, and experience?

PURCHASING QUALITY

1) Has Purchasing established procurement policies?
2) Does the Quality Manager ensure that purchased products and services conform to quality requirements?
3) Do quality managers assist Purchasing in evaluating and selecting suppliers and contractors based on their ability to provide consistent quality?
4) At a minimum, are suppliers and contractors evaluated based on objective measures, including:
   a) Relevant experience?
   b) Performance history, quality, cost, delivery, and stakeholder responsiveness?
   c) Supplier audit?
5) Does senior management reserve the right to require all or selective requirements of this Quality Management System on product/service suppliers, contractors, and external providers?
6) Do Purchasing Orders and Professional Service Agreements clearly describe all necessary requirements to ensure compliance with quality requirements?
7) Do Purchasing documents at a minimum include:
   a) Approved supplier and bidder list requirements?
   b) Approved Purchasing documents and authorities?
   c) Cost, quality, and delivery requirements?
8) Does senior management selectively verify the quality of purchased products and services?
9) Does verifying involve inspection, testing, or auditing to determine conformance with policies, procedures and specifications?
10) Are deviations, variances, and nonconformances rooting cause corrected?

QUALITY AUDITING

1) Does the auditee use internal and external audits to improve quality?
2) Do quality audits assess:
   a) Test environment?
   b) Test equipment?
   c) Calibration and traceability?
   d) Test procedures?
   e) Training?
   f) Handling, storage, and segregation of materials?
   g) Safety?
   h) Internal quality controls?
   i) Storage and handling of test equipment?
   j) Sample retention procedures?
   k) Supervision abilities?
3) Is quality audit documentation complete, current, and accurate?
4) Is corrective action pursued if indicated by the quality audit?
5) Are these audited:
   a) Quality program?
   b) Operations?
   c) Systems?
   d) Manufacturing processes?
   e) Service processes?
   f) Products?
6) Are there part or full-time quality auditors?
7) Is an interdisciplinary audit team used for complex audits?
8) Are quality auditors systematically evaluated?
9) Does the audit team use checklists?
10) Are auditors properly trained?
11) Does the quality audit team follow a systematic and approved procedure?
12) Are quality audit recommendations acted upon?
13) Is there a periodic audit of auditing internal controls?
14) Does the audit report address the following:
   a) Customer's name?
   b) Auditee's name and area?
   c) Auditor's name?

d) Purpose audit?

e) Tests used?

f) Recommendations?

g) Conclusions?

h) Corrective action?

15) Are quality audits periodically reviewed for patterns or other unusual conditions?

16) Do quality audits have conclusions and action recommendations?

17) Does the quality group have the authority to stop internal operations until customers are satisfied with products and services?

18) Is there a formal procedure of retrieving material as a result of:

a) Safety concerns?

b) Product performance?

c) Regulatory action?

19) Is there a formal procedure of resolving problems quickly?

20) Are laboratory, test, and measurement?

21) Are audits conducted?

QUALITY CONTROL

1) Is companywide quality management pursued?
2) If yes, does it involve:
   a) Top management support?
   b) Prevention driven, statistical process control?
   c) Companywide continuous improvement focus?
   d) Supplier wide continuous improvement emphasis?
   e) Customer focus?
3) Are all major quality functions, systems, processes, and activities identified?
4) Are quality controls established for all activities throughout the auditee's organization?
5) Are quality controls documented in quality standards, procedures, engineering specifications, flowcharts, or other quality-related documentation?
6) Do the following internal controls exist:
   a) Quality teams?
   b) Quality supervisors/managers?
   c) Quality standards?
   d) Quality policies?
   e) Procedures?
   f) Work instructions?
   g) Budgets?
   h) Progress reports?
   i) Corrective action requests?
   j) Audit reports?
   k) Internal/external reports?
   l) Checklists?
   m) Questionnaires?
   n) Objectives/goals?
   o) Plans?
   p) Benchmarks?
   q) Schedules?
   r) Methods?
   s) Testing/inspection?
   t) Machines/equipment?

7) Do quality plans address quality controls?

8) Are quality targets set throughout the auditee's organization?

9) Is the concept of quality control, assurance, and management universally understood?

10) Is the concept of variation universally understood?

11) Is the effectiveness and efficiency of the quality controls monitored?

12) Do quality controls have a feedback and correction mechanism?

13) Is continuous improvement pursued?

14) Are chance and assignable causes identified?

QUALITY COSTS

1) Is a budget for quality-oriented activities developed?
2) Are quality improvement projects identified and prioritized?
3) Are quality budgets developed for the:
   a) Organization?
   b) Business unit?
   c) Plant?
   d) Department?
   e) Team?
4) Are quality budgets prepared by the group responsible for meeting them?
5) Are quality budgets used to punish?
6) Are quality budgets revised without input from those responsible for meeting them?
7) Are the following costs tracked:
   a) Cost of quality (COQ)?
   b) Standard costs?
   c) Actual costs?
   d) Improvement project costs?
   e) Life cycle costs?
   f) Product development costs?
   g) Benchmark costs?
   h) Value added costs?
   i) Inspection and test equipment?
   j) Laboratory operation?
   k) Service operations control?
   l) External approvals?
   m) Test and inspection analyses?
8) Is the cost of quality (COQ) calculated?
9) Do internal failure costs include:
   a) Scrap?
   b) Rework and repair?
   c) Internal supplier scrap?
   d) Sorting?
   e) Engineering changes?
   f) Purchase order changes?

    g) Corrective action?

10) Do internal failure costs include:

    a) Warranty field failures?

    b) Customer complaint analysis?

    c) Product liability?

    d) Field service?

    e) Quality Costs-Comparison and Analysis

11) Is any area quality cost too high in relation to another area?

12) Is COQ data used to improve process or product quality?

13) Are standard rates established for labor, material, and overhead?

14) Are direct and indirect labor and material costs identified?

15) Are overhead rates calculated based on machine use?

QUALITY DOCUMENTATION

1) Is there a hierarchy of quality documentation consisting of:
   a) Vision statement?
   b) Mission statement?
   c) Objectives?
   d) Goals?
   e) Plans?
   f) Policies?
   g) Standards?
   h) Procedures?
   i) Drawings?
   j) Specifications?
   k) Work instructions?
2) Are above data current and complete?
3) Are the latest international, national, and industry methods of testing and measuring available, including:
   a) International Organization for Standardization (ISO)?
   b) American Society for Testing and Materials (ASTM)?
   c) American National Standards Institute (ANSI)?
   d) American Society for Quality Control (ASQ)?
   e) American Society of Mechanical Engineers (ASME)?
   f) Institute of Electrical and Electronic Engineers (IEEE)?
4) Are the following personnel records used and maintained?
   a) Organization chart
   b) Quality accountabilities
   c) Job descriptions
   d) Training
   e) Personnel policies and procedures
5) Are the following records used and maintained?
   a) List of equipment
   b) List of measuring equipment
   c) Equipment maintenance schedule
   d) Traceability records
   e) Test methods
   f) Sampling schedules
   g) Sampling methods

h) Internal audits
i) Calibration frequency records
j) Jigs and fixtures prints
k) Repeatability studies
l) Reproducibility studies
m) Inspection results
n) Laboratory test results
o) Product test results
p) Mechanical/physical/chemical tests
q) Reliability tests
r) QA standards
s) Specifications and drawings
t) Corrective action requests
u) Quality measurements
v) Process control charts
w) Product tests
x) Inspection results
y) Performance tests
z) Supplier quality standards
aa) Sales specifications
bb) Process capability analyses

6) Does quality documentation address:
   a) Performance?
   b) Cost?
   c) Workmanship?
   d) Time?
   e) Design?
   f) Delivery?

7) Is quality documentation:
   a) Valid?
   b) Uniform?
   c) Understandable?
   d) Doable?
   e) Current?
   f) Accurate?
   g) Available?
   h) Compatible?
   i) Complete?

8) Are changes incorporated into appropriate documents?

9) Are quality records retained?

10) Does the quality group review all engineering, purchasing, and manufacturing quality related documentation?

11) Is quality documentation retained in a safe location?

12) Does the auditee have a procedure that addresses the following in engineering?

13) drawing changes:
    a) Receipt?
    b) Approval?
    c) Distribution?
    d) Review of changes?
    e) Numbering?
    f) Disposal?

14) Do operating personnel have real time quality data on processes?

15) Do supervisory or lead personnel have access to quality management and customer survey reports?

16) Do quality reports track the following:
    a) Cost of quality?
    b) Prevention?
    c) Appraisal?
    d) Internal failure?
    e) External failure?
    f) Improvement projects?
    g) Customer satisfaction?
    h) Process capability?

17) Can the auditee's personnel obtain quality information quickly?

18) Is the company compiling quality related information into workmanship, specification, procedural, or quality manuals?

19) Is quality documentation periodically reviewed, revised, and approved?

20) Is a review date assigned to each quality document?

21) Do internal customers find reports timely, useful, and accurate?

QUALITY GOALS AND OBJECTIVES

1) Is there a:
   a) Quality vision statement?
   b) Quality mission statement?
2) Were all the following stakeholders solicited in developing the above:
   a) Employees?
   b) Management?
   c) Shareholders?
   d) Suppliers?
   e) Customers?
   f) Community?
   g) Government/regulatory agencies?
3) Have the following been developed:
   a) Quality objectives?
   b) Quality goals?
   c) Quality accountabilities?
   d) Quality plans?
4) Are these defined and developed for every organizational element?
5) Have quality objectives, goals, accountabilities, and plans been developed by key suppliers?
6) Does customer satisfaction drive company goals?
7) Are quality goals developed for:
   a) Ten years?
   b) Five years?
   c) Two years?
   d) One year?
8) Have the above goals been achieved?
9) Does the auditee communicate company goals so they are understandable to everyone?
10) Are quality objectives developed jointly by those who have to implement objectives?
11) Are corporate officers and other managers motivated through quality improvement?
12) Is quality performance measured companywide?
13) Are quality objectives and goals developed for:
    a) Business unit?

b) Plant?

c) Department?

d) Team?

14) Are quality goals and objectives actively pursued by middle management?

15) Are these quality goals and objectives actively pursued by workers?

16) Does the auditee communicate quality goals and objectives in terms of customer satisfaction, competitiveness, and profitability?

17) Are quality accountabilities developed for:

a) Chief executive officer?

b) Officers?

c) Managers?

d) Supervisors?

e) Workers?

18) Are accountabilities doable and measurable?

19) Are quality benchmarks established?

20) Are benchmark data used to support and improve operations?

21) Are quality benchmarks based on what the customer expects, wants, and needs?

22) Is there a formal process of bench marking?

a) Best products

b) Best processes

c) Best practices

d) Best services

23) Are quality benchmarks developed by people responsible for meeting them?

24) Are benchmarks attainable, and have e they been attained in the past?

25) Are benchmarks established at:

a) Corporate level?

b) Business unit level?

c) Plant?

d) Department?

26) Is competitive information gathered from:

a) Customer focus groups?

b) Dissatisfied customers?

c) Competitors?

d) Publications?

e) Product analysis?

f) Internal operations?

g) Supplier performance?

27) Are data used to continuously improve?

QUALITY INFORMATION AND ANALYSIS

1) Does the auditee identify its quality data needs?
2) Are the best ways to obtain data identified?
3) Are time requirements to obtain and analyze data considered?
4) Is data acquisition automated?
5) Are internal or external customers satisfied with data gathering, transfer, and analysis?
6) Is internal customer satisfaction measured and monitored?
7) Are quality data collected from:
   a) Customers?
   b) Suppliers?
   c) Employees?
   d) Internal operations and processes?
   e) Safety, health, and regulatory authorities?
8) Is quality information gathered at the:
   a) Corporate level?
   b) Business unit?
   c) Plant?
   d) Department?
   e) Process?
   f) Machine?
   g) Assembly?
   h) Component?
9) Are quality costs identified and broken down into:
   a) Internal costs?
   b) External costs?
   c) Appraisal costs?
   d) Prevention costs?
10) Are quality costs used at the:
   a) Corporate level?
   b) Business unit?
   c) Plant?
   d) Department?
   e) Team?
   f) Process?
   g) Machine?

11) Is there a systematic effort to lower or balance various COQ elements?

12) Do prevention costs include:
   a) Quality engineering?
   b) Reliability engineering?
   c) Preventive maintenance?
   d) Product design control engineering?
   e) Process control engineering?
   f) Quality measurement and inspection equipment?
   g) Prevention personnel?

13) Do appraisal costs include:
   a) Inspection and test setup?
   b) External supplier material control?
   c) Internal supplier material control?
   d) Inspection and test equipment?
   e) Laboratory operation?
   f) Service operations control?
   g) External approvals?
   h) Test and inspection analyses?

14) Do internal failure costs include:
   a) Scrap?
   b) Rework and repair?
   c) Internal supplier scrap?
   d) Sorting?
   e) Engineering changes?
   f) Purchase order changes?
   g) Corrective action?

15) Do internal failure costs include:
   a) Warranty field failures?
   b) Customer complaint analysis?
   c) Product liability?
   d) Field service?
   e) Quality Costs-Comparison and Analysis?

16) Is any area quality cost too high in relation to another area?

17) Is COQ data used to improve process or product quality?

18) Are standard rates established for labor, material, and overhead?

19) Are direct and indirect labor and material costs identified?

20) Are overhead rates calculated based on:
   a) Machine use?

b) Output?

c) Direct material costs?

21) Are standard costs derived through:

a) Time and motion studies?

b) Process analysis?

c) Machine productivity?

d) Labor efficiency?

22) Are overhead costs distributed in a rational manner?

23) Is the distribution of costs reviewed periodically?

24) Are labor and material costs accumulated in a work order or similar documentation?

a) Does the documentation include:

b) Operations performed?

c) Operator's name?

d) Labor content?

e) Material content?

25) Are actual hours and standard hours compared periodically?

26) Is an acceptable range of actual hours compared to standard labor hours?

27) Is there a system to identify and address variances between actual and standard labor hours?

28) Are material usage reports generated:

a) Daily?

b) Weekly?

c) Monthly

29) Is actual material used compared to projected material use?

30) Are studies conducted to:

a) Reduce scrap?

b) Reduce labor overtime?

c) Improve quality?

d) Approve equipment purchases?

e) Improve productivity?

f) Reduce overhead?

31) Do overhead studies address:

a) Developing accurate and timely overhead costs?

b) Surveying plants to analyze indirect factory overhead?

c) Estimating total indirect factory overhead?

   d) Are the following addressed in comparing actual to standard costs: Comparisons analyzed weekly or monthly?

   e) Acceptable ranges established?

   f) Variances addressed?

32) Are comparisons between actual and standard costs maintained by:

   a) Part number?

   b) Work center?

   c) Department?

   d) Commodity?

   e) Plant?

QUALITY MANAGEMENT PRINCIPLES

1) Are the following total quality management principles followed:
   a) Total customer satisfaction is the goal of all stakeholders?
   b) The "voice of the customer" is internalized?
   c) Continuous improvement is a basic premise to TQM?
   d) Continuous improvement is company and supplier wide?
   e) Auditing facilitates continuous improvement?
   f) TQM relies on self and group initiative?
   g) Benchmarks and goals are established in all activities?
   h) Everything is measured?
   i) Prevention is pursued company and supplier wide?
   j) Flexibility is required to cope with change?
   k) All types of waste are eliminated?
   l) All employees are empowered to improve?
   m) Employees are cross-trained and compensated for learning?

QUALITY MANAGEMENT SYSTEM

1) Does a formal Quality system exist?
2) Is it documented?
3) Have the drivers of the quality system been identified?
4) Are they understood and accepted by the organization?
5) Does the Quality System focus on stakeholder/customer service?
6) What are the formal goals of the QMS?
7) Do they include:
   a) Achieve high stakeholder/customer service?
   b) Achieve the optimum level of quality for each invested dollar?
   c) Exercise responsible care and due diligence?
   d) Scale quality functions to meet programmatic, project, and activity level requirements?
   e) Achieve adherence to standards?
   f) Maintain programmatic, project consistency, and quality?
   g) Establish and maintain quality checks and balances?
   h) Emphasize programmatic and project coordination, communication, collaboration, and conflict resolution?
   i) Enhance institutional knowledge and memory?
8) Is Director and ES Director authorized to form a Quality Management System?
9) The Quality Management System consist of 3 levels:
   a) Quality Management System at the programmatic level
   b) Quality assurance processes at the project level
   c) Quality control surveillance/test/inspection at the activity level

QUALITY MANAGEMENT SYSTEM GOALS

1) Does the QMS address the following goals:
   a) Increase and ES stakeholder confidence that programmatic, project, and activity levels are quality managed?
   b) Develop a clear and understandable framework for project managing and deploying quality?
   c) Describe a Quality Management System to stakeholders, Director and ES Director that makes sense?
   d) Provide sure-proof means for sharing information and collaborative decision making to and ES stakeholders?
   e) Assist project managers to understand their quality roles, authorities, and responsibilities?
   f) Increase oversight and defect-proof processes through closed, corrective action feedback loops?
   g) Move quality up the value chain from production inspection to design error prevention?
   h) Provide a clear and efficient visual approach to project manage quality?
   i) Provide checks and balances throughout?
   j) Ensure project consistency by documenting key quality processes?
   k) Provide quality guidance for key project activities?
   l) Document and ensure legal, code and other compliance?
   m) Is sufficiently flexible to accommodate command-control or collaborative project management?
   n) Provide clear framework of quality requirements to contractors and subcontractors?

QUALITY MANAGEMENT SYSTEM REVIEWS

1) Has senior management authorized a programmatic, Quality Management System review to determine:
   a) Effectiveness, adequacy, and suitability of the QMS?
   b) Value generation?
   c) Stakeholder satisfaction?
   d) Critical or major nonconformance to regulations, specifications, etc.?
   e) Critical assessment of risks?
   f) Quality trending of nonconformances and other critical quality data?
   g) Continual improvement through application of lessons learned?

2) Does the Quality Manager regularly review Quality Management System effectiveness at the programmatic, project, and activity levels?

3) Are continuing value, suitability, adequacy, and effectiveness of the QMS periodically assessed?

4) Does the Quality Management System reviews include periodic review of current quality performance and improvement opportunities related to:
   a) Project manager, contractor, designer, supplier, and other significant parties' performance?
   b) Stakeholder satisfaction?
   c) Quality audits/assessment results?
   d) Project quality conformance analysis?
   e) Status of outstanding preventive and corrective actions?
   f) Follow up actions from earlier management reviews?
   g) Changing programmatic, project, or people circumstances?
   h) Quality implications of new regulations or requirements?

5) Are Quality Management System review observations, recommendations and conclusion recorded to:
   a) Initiate corrective and/or prevention actions?
   b) Ensure nonconformances and variances are root cause corrected?
   c) Verify implementation of solutions?
   d) Ensure continual programmatic and project improvement?

6) Does the Quality Manager have the authority to change the scope, scalability, and application of the QMS?

7) Are waivers, concessions or reduction/expansion of scope of the QMS reviewed and approved by the Quality Manager?

QUALITY MANAGEMENT/ASSURANCE REVIEWS

1) Does the Quality Manager with consultation with functional and project management define, plan, and implement a closed-loop improvement process at the process or project level involving the following:
   a) Identify project quality processes and activities to monitor?
   b) Determine who will conduct the inspection or test?
   c) Determine type and location, timing, and frequency of measurement?
   d) Record the results?
   e) Issue CARs or PARs if required?
   f) Monitor the effectiveness of the CAR or PAR?

QUALITY CONTROL REVIEWS

1) Has the Quality Manager with consultation with project management identified areas of high risk, nonconformances, variation, or variances?
2) Based on this assessment, does quality management determine the need for test, inspection, certification, waiver, or release?
3) Is this information noted in the quality control form, which identify the following:
   a) Reasons for inspection/test?
   b) Required standards, specifications, requirements, codes, or regulatory issues?
   c) Inspection/test locations?
   d) Project characteristics, activities, products, etc. to be inspected/tested?
   e) People, procedure, equipment, and test critical to be used?
   f) Witness or verification if required?
   g) Final inspection to determine completion and effectiveness?
   h) Does the Quality Manager identify critical stakeholders and develop processes to measure/analyze/improve programmatic, project, and activity level quality performance?

QUALITY MANAGEMENT TEAM

1) Does the Quality Manager authorize the formation of a Quality Management Team?

2) Is the Quality Management Team is composed of Quality Manager (chair), Program Management Quality Team Leader, Design Quality Team Leader, and Production Quality Team Leader?

3) Does the Quality Management Team:
   a) Monitor stakeholder satisfaction?
   b) Issue and review and ES quality policies, procedures and work instructions?
   c) Authorize Quality Management System reviews, quality assurance audits, or quality control checks?
   d) Issue nonconformance findings?
   e) Review corrective actions?
   f) Review preventive actions?
   g) Resolve quality disputes?
   h) Issue waivers or concessions?
   i) Maintain master list of outstanding corrective actions and preventive actions?

4) Does the Quality Manager chair the Quality Management Team?

5) Does the Quality Manager report to senior management on the status of outstanding quality management issues that have high financial impact, significant potential on operations, risk, or public confidence?

QUALITY MANAGER

1) Is there a person who is the designated Quality Manager/Director?
2) Does this person conduct the following functions establish or assist in developing quality management quality policies and objectives?
3) Does this person ensure that program, project, activity level quality functions conform to critical stakeholder requirements?
4) Does this person have the authority to establish a programmatic level Quality Management System?
5) Have processes been established for satisfying key stakeholder requirements?
6) Does this person have the authority to ensure programmatic, project, and activity level consistency?
7) Does this person assist in programmatic and project risk mitigation?
8) Does this person have the authority to ensure compliance with regulatory, code and other requirements?
9) Does this person or another person authorize programmatic, project, and activity level quality reviews?
10) Does this person have the authority and scope of responsibilities to apply the QMS within and ES organizations and with supplier/contractors?
11) Does this person report to a senior corporate officer?
12) Have quality authorities been defined across functions, lines of authority, reporting relationships, interfacing, and communication flows?

QUALITY MANUAL

1) Is there a quality manual at the:
   a) Corporate level?
   b) Business unit level?
   c) Plant level?
   d) Department level?
   e) Work area/group level?
2) Does the quality manual:
   a) Outline a policy?
   b) State goals?
   c) State an objective?
   d) Define a procedure?
   e) State quality accountabilities?
3) Does the manual address:
   a) Supplied material control?
   b) Supplier selection/certification?
   c) Supplier monitoring?
   d) Supplier improvement?
   e) Material testing, inspection, and examination?
   f) In-process control?
   g) Statistical process control?
   h) Training?
   i) Sampling plans?
   j) Up-to-date quality standards, specifications, and prints?
   k) Measurement calibration and standardization?
   l) Final product control?
   m) Packaging and labeling?
   n) Shelf-life control?
   o) Product handling?
   p) Product storage?
   q) Nonconforming material disposition?
   r) Proper segregation?
   s) Recall procedure?
   t) Corrective action?
   u) Customer satisfaction?
4) Is the manual complete, accurate, and up to date?

5) Are all projects, systems, processes, procedures, and work instructions in the quality manual followed?
6) Does the quality manual properly de scribe all the quality functions?
7) Is there a written method for adding, deleting, and revising quality procedures?
8) Are methods established for determining internal and external customer requirements?

QUALITY MEASUREMENT

1) Are service and products quality levels compared against others?
2) Are these comparisons conducted on:
   a) Principal competitors?
   b) Industry averages?
   c) Industry leaders?
   d) World leaders?
3) Are comparisons based on:
   a) Independent surveys?
   b) Laboratory tests?
   c) Benchmarks?
   d) Company tests?
4) Are results compared against:
   a) Accuracy?
   b) Reliability?
   c) Timeliness?
   d) Performance?
   e) Behavior?
   f) Delivery?
   g) After-sales service?
   h) Documentation?
   i) Appearance?
   j) Durability?
   k) Maintainability?
5) Are results evaluated on:
   a) Improvement?
   b) Efficiency?
   c) Effectiveness?
   d) Economy?
   e) Existence?
   f) Occurrence?
   g) Completeness?
   h) Accuracy?
   i) Precision?
   j) Valuation?
   k) Disclosure?

l) Allocation?

m) Reasonableness?

n) Sufficiency?

o) Simplification?

p) Timing?

q) Validity?

r) Ownership?

s) Classification?

t) Conformance?

u) Performance?

6) Is the supplier quality evaluated in terms of the above factors?

7) Are companywide quality measurement systems developed and used?

8) Do measurement systems focus on continuous improvement?

9) Are these quality indicators used:

a) Customer satisfaction?

b) Market share?

c) Cost of quality?

d) Defect levels?

QUALITY ORGANIZATION

1) Is a corporate-level quality group required?
2) Are quality responsibilities and authorities fully operationalized?
3) Are quality objectives, goals, and ac accountabilities assigned to each worker?
4) Is quality a part of everyone's job?
5) Does the quality group have corporate visibility and authority?
6) Is the quality group fully independent?
7) Does the quality group report to top management?
8) Does the quality group have a well-defined mission and vision?
9) Does the auditee have a formal organizational chart?
10) Is the role of the quality group well defined?
11) Is the quality group responsible for:
    a) Quality training?
    b) Quality manual?
    c) Auditing?
    d) Measurement?
    e) Inspection/testing?
    f) Quality specification and standard evaluating?
12) Are quality accountabilities and authorities assigned as far down the organization as possible?
13) Do managers and workers feel they have sufficient information relating to quality?
14) Do these quality systems exist?
    a) Management system?
    b) Cost containment system?
    c) Service/delivery system?
    d) Manufacturing system?
    e) Engineering system?
    f) Supplier quality system?
15) Does the quality program comply with:
16) Customer's most stringent quality requirements:
    a) MIL-Q-9858A?
    b) ISO 9000?
    c) Other national/international quality standards?

QUALITY PLANNING (ASSURANCE)

1) Do operating personnel, project managers and quality engineers identify and plan the activities needed to achieve project level, quality assurance?

2) Are quality assurance plans consistent with other requirements of the Quality Management System and the results documented and submitted to the Quality Management Team?

3) Does quality assurance planning ensure that process and project changes are controlled and the intent of the Quality Management System is properly maintained during this change?

4) At a minimum, does quality assurance planning:
   a) Identify potential quality problems at the project level?
   b) Identify quality processes required to comply with programmatic level quality policies?
   c) Develop quality procedures to minimize risks?
   d) Identify and obtain sufficient resources to implement QA plans?
   e) Identify project quality activities to achieve the desired quality results and assurance?
   f) Verify quality results activities through audits, testing, or inspection?
   g) Review lessons learned?
   h) Submit quality plans to Quality Management Team for review and approval?

QUALITY PLANNING (CONTROL)

1) Do project managers and quality engineers identify and plan the activities needed to achieve activity level, quality control?
2) Are quality control plans consistent with other requirements of the Quality Management System?
3) Are the results documented and submitted to the Quality Management Team?
4) Does quality control planning ensure that key project activities are measured and tested to ensure conformance with work procedures, test instructions, and etc.?
5) At a minimum, does quality control planning:
   a) Identify potential quality problems at the project activity level?
   b) Identify critical quality control (inspection/test) points?
   c) Identify specific production activities to be inspected or tested?
   d) Identify acceptance criteria, special tools, and certified test personnel?
   e) Identify critical standards and specifications?
   f) Identify controls to minimize risks or nonconformances?
   g) Develop work instructions and checklists for ensuring consistency?
   h) Obtain sufficient resources?
   i) Submit quality plans to Quality Management Team?

QUALITY PLANS (GENERAL)

1) Does the auditee use a formal quality planning process?
2) Is quality improvement planning integrated into overall business planning?
3) Does the strategic quality plan incorporate:
   a) Customer orientation?
   b) Quality products and services identified?
   c) Standards identified?
   d) Realistic and measurable standards?
   e) Defect prevention?
   f) Continuous improvement?
4) Are strategic quality plans developed?
5) Are strategic quality plans implemented?
6) Are tactical quality plans developed?
7) Are tactical quality plans implemented?
8) Are quality plans developed for:
   a) Organizational planning?
   b) Companywide quality improvement?
   c) Supplier wide quality improvement?
   d) Quality costing?
   e) Training and development?
   f) Accountabilities?
   g) Management reviews?
   h) Problem solving?
   i) Corrective action?
   j) Product development?
   k) New products?
   l) New processes?
9) Does quality planning involve:
   a) Quality characteristic identification?
   b) Personnel requirements?
   c) Test equipment requirements?
   d) Control plan development through product lifecycle?
   e) Manufacturability studies?
   f) Process control analysis?
   g) Capability analysis?

    h) Material control?

    i) Statistical planning?

10) Is quality planning pursued at the:

    a) Corporate level?

    b) Business unit?

    c) Plant?

    d) Department?

    e) Work teams?

11) Are responsibilities for quality planning assigned to teams or to company officers?

12) Are front-line workers or teams consulted on quality plans?

13) Do front-line workers or teams develop their own quality plans?

14) Are these used in developing plans?

    a) Customer requirements

    b) Process capabilities

    c) Competitive/benchmark data

    d) Supplier capabilities

15) Are quality plans and goals deployed to all work units and suppliers?

16) Is performance relative to plans and goals periodically reviewed and acted upon?

17) Are quality plans based on current and future quality requirements in target markets?

18) Are results used to develop better plans?

19) Are quality plans prioritized?

20) Are quality plans periodically reviewed?

21) Is corrective action pursued on the quality plans?

22) Do quality plans incorporate internal quality controls?

23) Are new product, contract, or process plans identified and investigated for quality?

24) Are quality plans complete and extensive?

QUALITY RESULTS

1) Does the Quality Manager have the authority to ensure that quality plans are developed at the programmatic, project and activity levels?
2) Do quality plans focus on ensuring that the results of design or production, and other key quality activities satisfy quality requirements?
3) Do quality plans at all levels identify:
   a) Key stakeholders and their requirements?
   b) Applicable requirements and codes?
   c) Sequence of activities to ensure quality conformance and improvement?
   d) Appropriate interfaces and interactions?
   e) Audits if required?
   f) Corrective Action Requests and Prevention Action Requests?
4) Does the Quality Manager ensure quality processes are controlled and satisfy stakeholder requirements?
5) Does quality management establish quality methods and practices to achieve process and project consistency, including:
   a) Implement quality criteria and methods to control project quality processes, achieve project consistency and ensure conformance with organizational requirements?
   b) Verify project processes comply with organizational quality policies and procedures?
   c) Measure key project quality variables, monitor project quality processes, and follow up to ensure quality requirements are satisfied?
   d) Ensure key quality information and data are collected, reviewed and if necessary communicated to senior management?
   e) Maintain proper design, production, quality records to provide evidence of process, programmatic, and project quality management?

QUALITY WORKPLACE

1) Is the work environment conducive to the well-being and growth of all employees?
2) Is there a companywide quality aware ness program?
3) Is there a supplier wide quality aware ness program?
4) Are quality awareness efforts effective and believable?
5) Does the auditee have a good reputation for:
   a) Quality workplace?
   b) Quality of work life?
   c) Good place to work?
   d) Fairness?
6) Does the auditee have a good reputation for:
   a) Quality products?
   b) Quality service?
   c) Cost competitiveness?
   d) On-time delivery?
   e) Innovation?
   f) Continuous improvement?
7) Is goodwill and fairness established with:
   a) Community?
   b) Suppliers?
   c) Management?
   d) Employees?
   e) Shareholders?
   f) Minorities?
   g) Governmental bodies?
8) Are there clear statements of the auditee's quality vision and mission?
9) Are the following factors considered in improving employee well-being and morale?
   a) Health and safety
   b) Satisfaction
   c) Ergonomics
   d) Child care
   e) Counseling
   f) Recreational facilities
   g) Cultural understanding facilities

    h) Non-work-related education

10) Are these trends and key indicators monitored?
    a) Safety
    b) Absenteeism
    c) Turnover
    d) Attrition rate of customer contact personnel
    e) Satisfaction
    f) Strikes
    g) Worker compensation

11) Does the company ensure:
    a) Security?
    b) Fairness?
    c) Equal treatment?

12) Is the physical work environment monitored for:
    a) Temperature?
    b) Humidity?
    c) Circulation?
    d) Contaminants?

RISK MANAGEMENT - EMERGENCY PLANNING

1) Are emergency plans developed for?
   a) Earthquakes
   b) Tornadoes
   c) Chemical explosions
   d) Waste spills
   e) Floods
   f) Fries
   g) Toxic fume emissions
   h) Security breaches
2) Are employees trained about the responsibilities in the event of an emergency?
3) Are there regularly scheduled drills for different eventualities?
4) Is automatic emergency monitoring and shutdown equipment installed?

SALES/MARKETING STRATEGY

1) Are stakeholder/customer requirements addressed?
2) Are stakeholders needs and wants identified?
3) Are quantitative or qualitative assessments used to identity stakeholder/customer requirements?
4) Are stakeholder and particularly customer interviewed regarding requirements?
5) Are focus groups conducted?
6) Are stakeholder/customer surveys conducted?
7) What are the marketing and customer drivers for the organization?
8) Are customer purchasing behaviors monitored and quantified for various customer segments:
   a) Industrial
   b) Commercial
   c) Customer
9) Is customer satisfaction monitored for products as well as services?
10) Is customer satisfaction for the various market segments quantified?
11) Is satisfaction monitored throughout the customer-product lifecycle?
12) Is there a compliant identification and resolution process?
13) Is there a formal feedback process back to the complainant responding and correcting the complaint?
14) Are changes in market and customer expectations monitored?
15) Is this broken down by product segment?
16) Are the strengths and weakness of product and service offerings?
17) Is horizon or market planning conducted to determine future customer requirements?
18) Are new innovations, refinements, or offerings identified for meeting present and future requirements?
19) Are competitive requirements and offerings identified?

STRATEGIC MANAGEMENT

1) Is the external environment, risks, and opportunities monitored?
2) Are the compelling product and service benefits identified?
3) Is the competition identified by product and service category?
4) Are competitive factors understood and analyzed?
5) Are the following macro and micro trends identified, understood, and analyzed?
   a) Economic
   b) Cultural
   c) Technological
   d) Political
   e) Regulatory
   f) Social
   g) Demographic
   h) Ecological
6) Is the company's business model and concept articulated and well understood?
7) Are there an overall organizational strategy, plan, and tactics for reaching the business goals?
8) Are similar strategies, plans, and tactics for business units?
9) Is there a long-term vision?
10) Are operational goals aligned with overall business goals?
11) Are the following considered in the business environment planning?
   a) Regulatory compliance
   b) Training and educating employees
   c) Management development
   d) Pollution prevention
   e) Environment remediation
   f) Emergency response
   g) Federal, state, and governmental relations
   h) Public relations
   i) Merger, acquisition, and divestiture
   j) Environmental management

SUPPLIERS/CONTRACTOR QUALITY

1) Has senior management authorized policies to select, monitor, and improve key suppliers/contractors/vendors in conjunction with Purchasing?
2) Is Purchasing responsible for establishing purchasing policies?
3) Is the Quality Manager responsible that quality policies are consistently followed?
4) Do the Quality Manager and Purchasing Manager:
    a) Develop quality acceptance criteria for selecting key suppliers?
    b) Evaluate supplier's quality capabilities through auditing or inspection?
    c) Monitor supplier quality performance?
    d) Move from product inspection to defect prevention?

SUPPLIER PARTNERING

1) Is a large percentage of the sales and manufacturing dollar outsourced?
2) Is the make or buy decision formally analyzed?
3) Are the following considered in the make or buy decision?
   a) Total cost of alternatives
   b) Return on investment
   c) Available capacity
   d) Delivery
   e) Service
   f) Training
   g) Inventory and storage costs
   h) Quality capability
   i) Technical content
   j) Plant locations
   k) Production capacity
4) Does the customer have a formal supplier-partner quality management program?
5) Are suppliers single sourced?
6) Do the customer's internal departments work together to develop mutually agreeable plans to interface with suppliers?
7) Are examples available of implementing internal continuous improvement and customer satisfaction?
8) Do supplier-partners have a quality program for purchased:
   a) Production materials?
   b) Components?
   c) Subassemblies/assemblies?
   d) Services?
9) Are continuous improvement and other requirements spelled out to supplier-partners?
10) Are most suppliers considered cooperative?
11) Are supplier-partners' products and services recognized as high quality?
12) Does supplier-partner management communicate continuous improvement, prevention, and customer satisfaction requirements internally?

13) Are quality principles high priorities to supplier-partner management?

14) Does supplier-partner management at tend seminars on quality and statistics?

15) Does supplier-partner management encourage participation and involvement?

16) Does top supplier-partner management understand concepts of variation, statistical control, process capability, and continuous improvement?

17) Does supplier-partner management follow a disciplined approach to quality development and planning?

18) Does supplier-partner management follow a strategy for implementing statistical control and pursuing continuous improvement?

19) Does supplier-partner management identify and eliminate inhibitors to continuous improvement?

20) Have supplier-partners developed the following:
   a) Quality function deployment (QFD)?
   b) Control plans?
   c) Process flowcharts?
   d) Feasibility analyses?
   e) Failure mode and effects analysis?
   f) (FMEA)?
   g) Packaging/labeling/storage/handling plans?
   h) Preventive maintenance?

21) Is there a quality program for after-market sales?

22) Do supplier-partners have a companywide defect-prevention program?

23) Do supplier-partners make recommendations to the customer to improve the quality of supplied products and services?

24) Are supplier-partners concerned with and demonstrate:
   a) Quality of work life?
   b) Safety?
   c) Affirmative action?
   d) Continuous improvement?
   e) Prevention?
   f) Just in time?
   g) Training and development?
   h) Employee empowerment?

SUPPLIER TRAINING

1) Do suppliers have a formal training plan including a timing chart to satisfy customer requirements?
2) Do suppliers have a statistical specialist or consultant available to implement a statistical prevention program?
3) Have suppliers developed training manuals and guidelines for statistical quality training?
4) Have all personnel been trained in statistical technique?

Index

A

accuracy, 98, 114, 159
American National Standards
  Institute, 138
American Society for Testing
  and Materials, 138
appraisal, 140, 144
auditee, 8, 13, 14, 29, 30, 58, 60,
  61, 62, 63, 99, 108, 132, 134,
  135, 140, 141, 142, 144, 161,
  164, 167
auditing, 13, 14, 62, 148, 161

B

bar coding, 84, 86
bills of material, 27, 28
business plan, 67
business unit level, 7, 142, 157

C

calibration status, 74
cause and effect diagram, 59
complaints, 14
computer modeling, 50
configuration controls, 27
contingency plans, 48, 50
continual improvement, 151
continuous improvement, 7, 8, 9,
  10, 111, 134, 135, 148, 160,
  173, 174
contract reviews, 24
control, 58, 109, 127, 128, 164,
  174
control system, 43, 84
corrective action, 166
corrective action procedures, 10
corrective action requests, 11

Cost of Quality, 14, 20, 136, 140,
  160
customer complaint analysis, 13,
  137, 145
customer complaints, 9, 118
customer interviews, 13
customer metrics, 12
customer requirements, 12, 13,
  14, 22, 31, 40, 42, 86, 87, 114,
  158, 170, 175
customer satisfaction, 12, 13, 16,
  18, 119, 141, 142, 144, 148,
  170, 173
customer satisfaction, 14, 140,
  157, 160
customer satisfaction, 13
customer service personnel, 18
customer-supplier
  communications, 22

D

dealer visit reports, 13
defect levels, 160
defective materials, 118
defective products, 81
design change control, 26
design changes, 26
design characteristics, 113
Design Of Experiments, 59, 108
design process, 36, 37, 38, 105
design quality, 38, 155
design quality, 35
design quality outputs, 37
design quality plans, 35
design quality review, 38
design requirements, 32
design review, 34, 35, 38, 40
design risks, 38
distribution, 83
document control, 43

drawing control, 32, 38

E

emergency plans, 169
employee development, 58
employee recognition, 18, 63
employee suggestions, 45
energy conservation, 45, 46
energy management, 45
engineering change number, 96
engineering changes, 136, 145
engineering control board, 27
engineering drawings, 28
engineering prints, 23, 42, 75,
    103, 106, 127
engineering specifications, 42
engineers, 29, 70, 138
environmental conditions, 25,
    112, 118
environmental policy, 49
environmental product testing,
    50
environmental recovery and
    waste, 51
environmental regulations, 47
environmental site selection, 53
external costs, 144
external failure, 140

F

failure mode analysis, 32
Failure Mode and Effects
    Analysis, 112, 128, 174
feedback mechanism, 18
field failure, 13, 118
field failure reports, 13
field service, 33, 137, 145
final inspection, 154
financial and physical resources,
    57
financial investment, 55
finished materials, 86
FMEA, 105, 113, 128, 174

G

gage capability, 96
gage certification documents, 96
geometric dimensional
    tolerancing, 23, 28, 72
Geometric Dimensional
    Tolerancing, 59

H

handling, 18, 22, 90, 94, 132
handling material, 22
hazardous materials, 47, 48, 50,
    51, 91
human resource management,
    62
human resource requirements,
    64
human rights, 66

I

improve productivity, 146
incoming shipments, 72, 73, 91
information and analysis, 144
information technology, 67
information technology control,
    69
in-process control, 106, 157
inspection, 74, 76, 77, 82, 98,
    106, 109, 123, 136, 139, 145,
    154, 161
inspection and testing of
    supplied products, 72
internal audits, 139
internal costs, 144
internal failure, 140
internal housekeeping, 116
international organization for
    standardization, 23
inventory system, 86
ISO, 23, 25, 66, 115, 138, 161
IT software development, 70

L

labels, 84, 96, 119
labor management relations, 65
legal and regulatory compliance, 78
lessons learned policy, 79
life cycle costs, 136

M

manufacturing, 10, 18, 28, 38, 47, 48, 52, 75, 86, 87, 97, 104, 106, 120, 124, 129, 140, 173
manufacturing and production, 120
manufacturing engineering, 33
manufacturing facilities, 10
manufacturing process documentation, 106
manufacturing work environment, 116
marketing and selling, 17
material disposition procedures, 81
material movement and handling system, 86
material quality, 90
material shipping procedures, 91
measurement, 28, 33, 43, 59, 63, 75, 79, 96, 98, 99, 100, 103, 114, 118, 124, 127, 128, 133, 145, 153, 160
measurement, 72, 76, 77, 98, 100, 109, 124, 157, 161
measurement and test equipment, 96
measurement equipment control, 100
measurements, 23, 29, 75, 79, 80, 100, 127, 139
MRP, 104

N

national standards, 23
new product and service design, 31
noncompliances, 47
nonconformance reports, 11, 102
nonconformances, 9, 11, 79, 90, 101, 102, 131, 151, 154, 163
nonconforming, 9, 106, 157
nonconforming products, 81, 82, 104, 105

O

objectives/goals, 134
operational baselines, 20

P

packaging, 88
packing materials, 88
Pareto analysis, 58, 108
part number, 28, 74, 88
patterns, 9, 12, 15, 53, 99, 113, 133
planning quality, 61
plant level, 7, 157
policies, 12, 22, 43, 49, 52, 53, 55, 66, 73, 79, 80, 101, 130, 131, 134, 138, 155, 156, 162, 166, 172
poor product performance, 118
precision, 98, 159
preproduction planning, 123
prevention, 101, 134, 140, 144, 145, 148, 166, 174
preventive action requests, 11, 102
preventive action requests, 102
preventive maintenance, 95, 103, 119, 125
preventive maintenance, 106, 145, 174

preventive maintenance, 125
principal competitors, 13, 159
procedures, 9, 12, 22, 43, 47,
    49, 51, 52, 53, 54, 55, 59, 66,
    70, 72, 73, 80, 83, 90, 93, 95,
    98, 101, 103, 106, 110, 120,
    128, 131, 132, 134, 138, 155,
    158, 162, 163, 166
procedures, 107, 134, 138
process capabilities, 165
process control, 7, 9, 106, 109,
    139, 145, 164
process controls, 7
process flow diagram, 105, 108
process instructions, 128
process level, 7
process monitoring, control, and
    improvement, 108
process variation, 7
product and workplace safety,
    117
product design and
    development, 25
product design control
    engineering, 145
product development, 32
product engineering, 33, 40
product knowledge, 18, 59
product labels, 84
product liability, 137, 145
product life cycle, 17, 25, 31
product packaging, 19, 88
product sampling and inspection
    procedures, 74
product storage procedures, 93
product test results, 139
product testing, 114
production and delivery of
    services, 115
production and manufacturing,
    104
production equipment, 31, 96
production materials, 122
production process, 31, 54, 108,
    126, 127

production quality, 155
production quality, 126
production quality review, 127
production schedules, 86
project characteristics, 154
project closeout, 103
project management resources,
    130
proper segregation, 81, 82, 157
proprietary products, 22
purchase order changes, 136,
    145
purchasing orders, 131
purchasing quality, 131

Q

QA standards, 139
quality, 7, 8, 9, 10, 13, 14, 18,
    20, 21, 22, 23, 26, 28, 29, 30,
    31, 35, 36, 37, 38, 42, 43, 44,
    53, 58, 61, 62, 63, 67, 69, 70,
    73, 74, 75, 76, 77, 78, 79, 80,
    86, 90, 100, 101, 103, 104,
    106, 108, 110, 111, 115, 116,
    118, 123, 124, 126, 127, 128,
    130, 131, 132, 133, 134, 135,
    136, 137, 138, 139, 140, 141,
    142, 144, 145, 146, 148, 149,
    150, 151, 153, 154, 155, 156,
    157, 158, 159, 160, 161, 162,
    163, 164, 165, 166, 167, 172,
    173, 174, 175
quality accountabilities, 138, 141
quality assurance, 11, 27, 79,
    102, 149
quality assurance planning, 162
quality attributes, 28, 73
quality auditing, 132
quality capability, 173
quality circles, 18, 63
quality control, 11, 43, 79, 102,
    149
quality control, 134
quality control planning, 163

quality costs, 137, 145
quality documentation, 23, 43,
   138, 139, 140
quality documentation, 138
quality engineering, 145
Quality Function Deployment,
   13, 59, 174
quality goals and objectives, 141
quality management, 11, 43, 69,
   79, 102, 130, 131, 149, 150,
   151, 155, 156, 162, 163
quality management system, 43,
   69, 102, 130, 131, 149, 150,
   151, 155, 156, 162, 163
quality management system,
   149, 150, 151
quality management system
   goals, 150
quality management system
   reviews, 151
quality manager, 11, 21, 26, 35,
   36, 38, 43, 69, 78, 79, 90, 101,
   102, 103, 126, 130, 131, 151,
   152, 153, 154, 155, 156, 166,
   172
quality manager, 156
quality manual, 161
quality manuals, 157
quality measurement, 159
quality performance, 61, 63, 141,
   151, 154, 172
quality plans, 164
quality program, 132
quality results, 166
quality standards, 106, 134
quality statistics, 58
quality system, 149
quality training, 161
quality workplace, 167

R

raw material, 42, 81
raw materials, 52, 83, 110, 122
raw materials, 9, 86

recall procedure, 157
reclaim, 81
recycling, 51, 52, 88
recycling, 52
reduce scrap, 146
regulatory approvals, 114
regulatory mandates, 93
reliability, 32, 120, 139, 145, 159
reliability engineering, 145
reliability testing, 112
reliability tests, 139
repeatability studies, 98
repeatability studies, 139
return on investment, 173
return to supplier, 81
rework, 81, 136, 145
risk assessment, 35, 101
risk management - emergency
   planning, 169
risk mitigation, 102, 156
risk/nonconformance control,
   101
risks, 15, 41, 47, 53, 57, 75, 79,
   90, 102, 114, 151, 162, 163,
   171

S

safety concerns, 118, 133
safety program, 120
sales, 139
sales, 12
salvage, 81
sampling methods, 138
sampling plans, 77, 157
sampling tables, 72, 75
scrap, 81, 136, 145
senior management, 43, 45, 61,
   103, 126, 130, 131, 151, 155,
   166, 172
service lifecycle, 115
shipping material, 23
social responsibility, 66
SPC, 23, 58, 73, 96, 108, 109,
   110, 123

specifications, 15, 22, 23, 28, 29, 30, 31, 38, 42, 70, 80, 87, 88, 90, 96, 101, 103, 106, 108, 110, 111, 123, 131, 134, 139, 151, 154, 157, 163
stakeholder complaints, 11, 102
stakeholder relationships, 20
stakeholder requirements, 20, 21, 36, 156, 166
stakeholder requirements, 21
stakeholders, 20, 21, 26, 35, 36, 38, 39, 45, 79, 103, 126, 141, 148, 150, 154, 166, 170
standard costs, 146, 147
Statistical Process Control, 157
strategic management, 171
strategic marketing, 170
supplier monitoring, 157
supplier partnering, 173
supplier training, 175
supplier-partner, 173, 174
suppliers, 29, 33, 40, 53, 58, 117, 141, 144, 167
suppliers/contractors quality, 172
supply management, 83
sustainability, 66, 121

T

Table Of Contents, 3
team performance, 63, 64
technology, 22, 31

test equipment, 132, 164
test product performance, 123
testing/inspection, 134
Total Quality Management principles, 148
toxic waste, 54
TQM, 20, 148
traceability, 81, 98, 114, 138
trained personnel, 124, 125, 127
transportation, 95
trends, 9, 13, 15, 113, 168, 171
type of deficiencies, 76

V

valuation, 159
value added costs, 136
value analysis, 39
variation, 7, 72, 96, 99, 135, 154, 174
verification, 35, 37, 38, 43, 70, 126, 154
voice of the customer, 148

W

waivers, 81, 152, 155
warranties, 56
warranty field failures, 137, 145
waste management, 54
work in process, 86
work instructions, 106, 134, 138

www.ingramcontent.com/pod-product-compliance
Lightning Source LLC
Chambersburg PA
CBHW081106220326

41598CB00038B/7252